Camp Logic:

A Week of Logic Games and Activities for Young People

Mark Saul and Sian Zelbo

ISBN: 978-0-9776939-6-2 (print)
ISBN: 978-0-9776939-7-9 (ebook)
Library of Congress Control Number: 2014954319

Camp Logic: A Week of Logic Games and Activities for Young People
By Mark Saul and Sian Zelbo
Illustrations by Sian Zelbo
Copy editor Carol Cross

Published by Delta Stream Media, an imprint of Natural Math
309 Silvercliff Trail, Cary, NC, USA

Natural Math

Logic Course

Overall Goals for this Course

The intent of this "course" is to introduce students to the very nature of mathematics. Mathematical truth is not empirical truth: it does not depend on experimentation and observation. Mathematical truth is not religious truth: it does not depend on faith or explore the ethics of human interactions.

Rather, mathematical truth is logical truth. What this means is that mathematics explores the relationships among statements and, in particular, which statements imply other statements.

The nature of implication as a logical relationship is subtle, and it will take students some years to fully appreciate it. However, central elements of the notion of implication are embedded in our intuition, and we can work to refine students' intuitions about these logical relationships.

In particular, younger students very naturally accept that new information can be derived from old: that if we are told certain things, certain other things follow. While a deeper description of this phenomenon may elude them, they can practice and sharpen their intuitions with interesting puzzles, stories, and games.

Many of these puzzles do not, on the surface, resemble traditional mathematics. It may come as a surprise to the students (and often even to their parents!) that in working these puzzles, they are practicing some of the same skills that they will be harnessing in learning more traditional mathematical material.

The notion that if statement A is true, then statement B is also true ("if A then B") is one that comes naturally to students. The subtleties come later, when we ask under what circumstances A does not imply B, when students must distinguish this statement from "if B then A," or when they explore the notion of indirect proof (proof by contradiction). We offer several examples in the problems and puzzles below.

Most of these problems are almost content free. That is, students need know very little mathematical background to work them. However, we have taken the opportunity of talking about logic to introduce some very specific logical principles that are often used in mathematics. These include the pigeonhole principle, invariants, parity, and isomorphisms. Again, these concepts are very subtle, and students will learn much more about them in advanced work. But an early introduction can only benefit them as they grow intellectually.

We have often included advanced notes for the teacher, which are not always appropriate for the student. It is important for the teacher to know more than he is trying to teach. (If a teacher is teaching addition, he should certainly know something about what it leads to—and even more than just multiplication.)

The material is organized so that it can be used in an after-school program or informal setting. As a stand-alone curriculum, such as might be used in a math summer camp, it will fill a full week. Some of the activities depend on previous activities, and we have tried to indicate pedagogical threads that weave the activities together. However, most of the activities can also be extracted and used independently, perhaps to fit into some other pedagogical structure.

Our point of view about elementary mathematics, including some of these amusing and simple problems, is that it must be seen as "embedded" in much more serious mathematics. Mathematics as a discipline has a certain unity: every small part of it betrays the structure of the whole. And that structure is its logical structure.

We hope your students enjoy solving these puzzles as much as we have enjoyed selecting them.

Acknowledgements

This book would not be possible without Maria Droujkova's organizational skill, vision, and constant support.

The core activities from this book were developed through work with the children of Bronxville, New York, and the math circles held there. We would especially like to thank the late Debby Rankin, at whose suggestion and with whose support these math circles were first organized.

The activities in this book were further developed and refined with the help of many people in math circles and in after-school and classroom settings all over New York City. This enormous undertaking would not have been possible without the support and feedback from the open-minded administrators and educators who embraced an unorthodox approach to math education. While we cannot name everyone, we would like to thank (in alphabetical order) Ethan Berman, Connie Burton, Nancy Burton, Candace Reyes-Dandrea, Phil DiTuri, Melissa Flores, Meghan Groome, Nadine Hibbert, Esther Kogan, Maggie Maluf, Lisa Mielke, Bob Reveri, Athena Shapiro, Vicky Vlantis, and Stephanie Wortel.

We also would like to thank Moumanti Podder, James Fennell, and Mike Pienciak for helping us prepare the manuscript.

The vast majority of this work was the result of a generous grant from the Alfred P. Sloan Foundation to the Center for Mathematical Talent, Courant Institute of Mathematical Sciences, New York University. This grant brought together people with a wide variety of expertise, resulting in this book.

Big thanks to our crowdfunders for their support in producing the book:

Anonymous
Adrian Smith
Albina
Alexandr Rozenfeld
Alexandra Fradkin
Alexey Yeltsov
Alyssa Mattingly
Amanda Serenevy
Andrea Gray
Andrea Landaker
Angela Smale
Angelique LeDoux
Anna Burago
Ashley Ahlin
Bernadette
Miragliotta
Brandy Wiegers
Brian Chess
Brian Lay
Cameron Goble
Carol Cross
Caroline Mukisa
Casey Christensen
Cate Cugell
Charla Mair
Chen Zhao
Christina
Christine Woskett
Christopher
Clea Danaan
Edelblute
Corin
Craig Provost
Cristina Cardone
Dale Provost
DaNel Hogan
Daniel Finkel
Daniel Zaharopol
Danielle
Dara Kubovy-Weiss
Denise Gaskins
Dor Abrahamson
Elena Koldertsova
Emily Smith
Eric Schreiber
Frank Eakin
Gangadhar
Gerald Ardito
Gleb Groysman
Glynns
Heather
Holly Brown
Holly Orcutt
Howard Zelbo
Hulya Eraslan
Ihor Charischak
Jack & Pam Stewart
Jaclyn Koehl
Japheth Wood
Jarkko Laine
Jason Holder
Jason Jarrett
Jelena
Jennifer
Jennifer Mendez
Jessica Christoforo
Jill Zwickl
Jim Burtoft
John Bibby
Jorrit de Groot
Joshua Zucker
Julia B.
Julie Sundaram
Jyoti
Karen Snowball
Kathy Cordeiro
Kathy Macgowan
Kathyann Natkie
Kyle Dean
L .
Laura
Laura Wickline
Lee Zee
Lesley Yeo
Lew Douglas
Libbie Rice
Linda Henderson
Linda Zanieski
Logan Hobley
Lori Philipson
Lucinda Turck
Luke Sinden
M. Spiker
Margaret Maluf
Martin Gannholm
Matthias Kawski

Maud Schlich
Maureen Tumenas
Mindy Ranney
Nancy Sedlacko
Nari Lewis
Naveen Srinivasan
Noah
Olga Lepekhina
Patricia Pacelli
Phuong Ngo
Pilar Alayo
Raamganesh Samakkulam
Rachel Flenley
Ravi Subramanian
Rebecca Bucci
Remy Poon
Renae Hwang
Richard W Hill
Robert Deskin
Saad Mneimneh
Sally Hinton
Santosh Zachariah
Scott Provost
Sergio Valenzuela
Shiying Lee
Sondra Papanek
Su Penn
Susan
Susan Barbour
Susan Million
Teresa Gonczy
Terri Coons
Tomas Fabian Gonzalez
Tracy Romzek
Vinci Daro
Vivian Figueredo
Willard H. Blaskopf Jr.
Yee-Wah Lee

A Word About Style

Throughout the book, the pronouns "he," "his," and "him" are used as gender-neutral pronouns.

Overview

Day One:
Animal Puzzles (An Introduction to Logical Reasoning with Cryptarithms)
The Game of Giotto (Practice with Pure Logical Reasoning)
Lewis Carroll Puzzles (Proof by Contradiction)
Cryptogram Puzzle (More Proof by Contradiction)

Day Two:
Giotto Puzzles (Analyzing the Logical Structure of Giotto)
Watermelon Language (Logic Applied to Number Systems)
Jittery Soldiers (An Introduction to Invariants)
The Black and Red Problem (An Introduction to Parity)

Day Three:
Parity Problem Set
Discussion of Black and Red Problem (A Surprising Connection to Parity)
Ginger's Pigeons (Proofs with the Pigeonhole Principle)
The Mouse-and-Cheese Problem (Using the Idea of an Isomorphism to Solve a Problem)

Day Four:
Nim (An Introduction to Mathematical Induction)
Two-Row Nim and One-Piece Chess (Another Example of an Isomorphism)
Leap Frog Activity (An Activity for Exploring Invariants)

Day Five:
The Boys and Girls Problem (Another Example of the Use of Invariants)
Hidden Cards Puzzle (Practice with Logical Deductions)
Magic Squares and 15 Game (More Practice with Invariants and Isomorphism)

Day 1

Goals for the Day

Students will be introduced to logic and the nature of "if-then" reasoning.
Students will also be introduced to a specific kind of "proof," proof by contradiction.

Animal Puzzles

We use a particular type of puzzle called a "cryptarithm" to introduce logical reasoning. The point is that while the puzzles look mathematical, the hard part here is the logic, not the arithmetic.

In a running math circle, it can work well to begin every day with a few cryptarithms. This will allow students to begin working on something familiar. The activity is "modular," so the timing is not well defined. For example, students who arrive early can work a few puzzles while waiting for everyone to settle in, or a math circle could use the whole set of puzzles as the day's activity. Finally, the puzzles do not always require sustained thought. Students can disengage and come back to them later on.

To begin, put puzzles on the board one or two at a time, and have the class work on them together. If students want to do more, many (difficult) examples can be found on the Internet.

As another extension, encourage the students to make their own puzzles. They are not easy to make. A classic cryptarithm is considered elegant if it has a unique solution. But for beginning students, the educational purpose is served if we relax this condition a bit.

For example, the cryptarithm AB + BC = CD has far too many solutions to be of interest, except perhaps as an exercise. On the other hand, the cryptarithm AB + BC = CA has no solutions at all. Again, this is not very satisfying as an introductory problem. But again, a proof that it has no solutions might make a good exercise for students later on.

Note, too, that these really are problems. The statement is easily comprehended. The goal is well defined. Yet there is no obvious algorithm, or sequence of steps, that will surely solve the problem (short of trying every value of the digits, which humans generally do not want to do). There is not even one pre-determined method of attack. These problems make students think.

The Rules

Each cryptarithm puzzle below represents an arithmetical statement. In other words, in a single puzzle a P might represent the digit 4, and if so, then it represents a 4 everywhere in the puzzle, and no other letter will have that value. But the value of P can be different in different puzzles. If we write AB, we mean the integer with tens digit A and units digit B. That is, juxtaposition of letters denotes place value, and not multiplication. Multiplication, either of single digits or multi-digit numbers, is indicated with a "×". We assume for this sort of puzzle that there are no leading zeroes (zeroes to the left of the numeral).

Introductory Puzzles

W + O = O F

Solution: We know that here a one-digit number plus a one-digit number gives us a two-digit number. The largest such sum is $9 + 9 = 18$. So O must be 1. (It cannot be larger.) If O $= 1$, then W $= 9$, because otherwise the sum would be one digit and not two. Answer: $9 + 1 = 10$.

P + P + P = I = G + G

Solution: The letter I must be a single-digit number divisible by 2 and 3. So it must be 6, which means P = 2 and G = 3.

Students may reason about it differently. Some will reach the same solution by plugging in numbers and ruling out possibilities. After they have solved the puzzle, it is useful to point out the more elegant solution we have given.

Notice that the PIG puzzle uses completely different reasoning from the previous puzzle.

Solution: The product A × A × A has only two digits. A quick check shows that the only candidates are 3 × 3 × 3 and 4 × 4 × 4, of which only 4 × 4 × 4 = 64 works (because BA must end with the digit A).

C + CA = ATT

Solution: The numbers on the left look too small to make a three-digit sum. The only way that the sum of this single-digit number and double-digit number can be a three-digit number is for C to be 9 and A to be 1. Answer: 9 + 91 = 100.

$$\begin{array}{r} ME \\ +ME \\ \hline BEE \end{array}$$

Solution: The sum of two double-digit numbers cannot be too large. A quick check (similar to the reasoning in the CAT puzzle) shows that B cannot be larger than 1, so we must have B = 1. Then E = 0 because E + E ends with E. So far we have M0 + M0 = 100. Because M0 is half of 100, M0 must be 50. Answer: 50 + 50 = 100.

HH + HH = OOT

Solution: As in the preceding problem, we have a hundreds digit that cannot be any larger than 1, so it must be equal to 1. We also know that HH + HH = 2 x HH = OOT, so OOT must be even. So T is an even digit, and the only options are 110, 112, 114, 116 and 118. Dividing each number by 2, we see that only 110 / 2 = 55 satisfies the given relation.

$$\begin{array}{r} GO \\ +\ ON \\ \hline ONO \end{array}$$

Solution: The number ONO is the three-digit sum of two double-digit numbers. As in previous problems, O cannot be larger than 1, so it must be equal to 1. Then N must equal zero because in the ones column, O + N = O. So far we have G1 + 10 = 101, so the solution is 91 + 10 = 101. (An "ono" is another name for a wahoo, which is a kind of fish.)

A + A + A + A + A + A + A + A + A + P = E

Solution: It is not hard to see that A cannot be too large, since nine A's plus a P is still a one-digit number. In fact, if A were as much as 2, the sum would be a two-digit number. So A = 0 or A = 1. If A were zero, then P and E would represent the same digit. So A = 1, and it follows that then E = 9, and P = 0.

Answer: 1 + 1 + 1 + 1 + 1 + 1 + 1 + 1 + 1 + 1 + 0 = 9

$$\begin{array}{r} \text{A A} \\ +\ \ \text{A} \\ \hline \text{E L K} \end{array}$$

Solution: In this problem A cannot be too small. In fact, if A were as small as 8, then AA + A could not be a three-digit number. So A = 9, and ELK is 108. Answer: 99 + 9 = 108.

```
  UKK
   KK
+   K
-----
  AUK
```

Solution: The letter K must be 5, because 3 × K gives us a K in the units place. The only other possibility for 3 × K = K is that K = 0, but that is not possible because K + K in the tens place gives a U and not a K. From there we can work out the rest easily. The U must be a 1 because of the carry from the ones place, so AUK = 215. Answer: 155 + 55 + 5 = 215. (An "auk" is a sea bird similar in appearance to a puffin.)

```
  T T F
+   T T
-------
  E F T
```

Solution: The F must equal zero because F + T = T. Then T must be 5, because T + T ends with a zero (in the tens place). Thus we have 55 + 55 = 605. (An "eft" is a life stage of a newt.)

What is the product of the following single-digit numbers?

$$C \times H \times I \times C \times K \times E \times N \times P \times O \times T \times P \times I \times E \times S$$

Solution: There are ten *different* digits, so one of them must be zero, and therefore the product is zero.

Challenge cryptarithm for those who may be working ahead of the class:

MA x CA = WWW

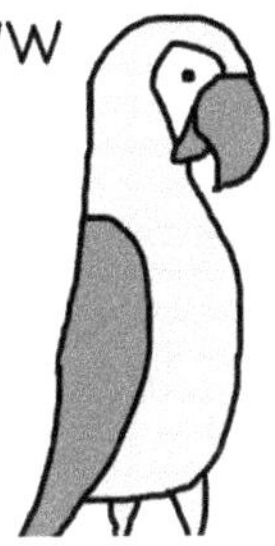

Solution: 37 × 27 = 999. WWW must be a multiple of 111. Because 111 factors into 3 × 37, we know that 37 must be a factor of WWW. Since 37 is prime and cannot be factored any further, 37 must be a factor of either MA or CA. In fact either MA or CA must be 37. (Experimenting a little will show that neither MA nor CA can be a higher multiple of 37, and so one of the two must equal 37.) Suppose MA = 37. Then the other factor (CA) cannot be too large, because the product is a three-digit number. In fact, 37 × 30 = 1110, which is too big, so C = 1 or 2. Since A = 7 and CA must be divisible by 3, CA can only be 27. Attentive students may note that the solution to this cryptarithm is technically not unique. We could have MA = 37, CA – 27, or vice versa. The cryptarithm is *symmetric* in these two numbers.

More Cryptarithms

LEVEL 1:

1. **AB - BA = A**

2. **A + BB = ADD**

3. **BC + B = DAD**

4. **BC + DC = CAB**

LEVEL 2:

5. **HEN + E = EGG**

6. **DAD + DAD + DAD + DAD + DAD = GLAD**

7. Find all possible solutions:

   AB + BA = (A+B)(A+B)

Solutions

1. **AB - BA = A**

The A must be a multiple of 9. (Play around with the difference of numbers of this form, and you'll see why.) In fact, since A is a single digit, it must be equal to 9. Answer: 98 – 89 = 9.

2. **A + BB = ADD**

The size of the numbers tells us that A must be 1. It is then not hard to see that the unique solution is 1 + 99 = 100.

3. **BC + B = DAD**

D is a carry from the tens place, so D = 1. Then it is clear that B must be a large digit, and in fact must be 9 (or the sum would not have three digits). We have 9C + 9 = 1A1, and we must have C + 9 = 11, so C = 2. Thus the unique solution is 92 + 9 = 101.

4. **BC + DC = CAB**

We know that C = 1, since the sum of two double-digit numbers cannot be bigger than 198. Then we see from the units column that B = C + C = 2. Now we have 21 + D1 = 1A2. Because we have a three-digit sum, D must be large; in fact it must be 8 or 9. But the choice of 9 gives A = 1, which repeats the value of C. Hence D must be 8. Answer: 21 + 81 = 102.

5. **HEN + E = EGG**

Since the hundreds digit in the sum is different from H, there must be a carry from the tens to the hundreds column. Likewise, since the tens digit in the sum is different from E, there must be a carry from the units to the tens column.

Examining the tens column, and considering the carries, we have E + 1 = G + 10, so E = G + 9. Since E is a single-digit number, we must have E = 9, G = 0. We now have H9N + 9 = 900. Answer: 891 + 9 = 900.

We have written the solution, as in other problems, using the language of algebra, which makes it easy for us to understand. Students may get the answer, but struggle to express themselves. At this stage, if students have the right answer, we can assume that they have an idea of the reasoning behind the answer and pull out of them whatever explanation they can offer. The ability to express oneself mathematically develops over time.

6. **DAD + DAD + DAD + DAD + DAD = GLAD**

Five times D ends with a D, so D must be 5 or 0. In fact it must be a 5 because we don't write leading zeroes (also because the sum is a four-digit number). Because 5 x 5 = 25, we have a carry of 2 from the units to the tens column.

From the tens column we see that 5 × A + 2 ends in A. That means A could be 2 or 7.

If A is 2, we have 525 + 525 + 525 + 525 + 525 = 2625, and G = A, which is not allowed.

If A is 7, we have 575 + 575 + 575 + 575 + 575 = 2875, which is the unique solution.

7. **AB + BA = (A+B) (A+B)**

When we add AB and BA, we get a multiple of 11, no matter what values A and B take on. We can see from the right side of the equation that the sum also gives us a square number, so it must be a square multiple of 11. The product on the right therefore must be 11 × 11. (It can't be any higher multiple than 11 itself since if it were as much as 22, we would have 22 x 22 = 484. But 484 cannot be the sum of two double-digit numbers.) The four solutions are:

29 + 92 = (2 + 9) (2 + 9)

38 + 83 = (3 + 8) (3 + 8)

47 + 74 = (4 + 7) (4 + 7)

56 + 65 = (5 + 6) (5 + 6)

The Game of Giotto

This is one of a number of puzzles and games drawing on mathematical linguistics. Students enjoy it, and it works well when played at the board with a whole class. Once students learn the game, it can be played in about twenty minutes, and so it can be used to fill little gaps of time at the end of a session. The game has many interesting aspects, but here we are using it to give students practice with logical reasoning and making that reasoning explicit.

Classroom Procedure

The game is best played on the board with the whole class participating.

The instructor should think of a secret five-letter word with no repeated letters. (Below we have chosen the word BREAD.) Write the secret word on the chalkboard in a place that can be covered up, or on a piece of paper that is folded and placed securely away from the students.

The students must work together to guess the word. They do so by offering their own "test" words, five letters each. You report to students the number of letters their word has in common with yours.

For example, suppose your secret word is BREAD and that students offer the following test words:

Test Word	Score	
NIGHT	0	
BRINK	2	
CHEEP	1	(Only one E counts in the score.)
DREAD	4	
BEARD	5	
BREAD	Right!	

(Note that you do not tell the students which letters the test word has in common with the secret word, nor anything about the order of the letters in the secret word.)

The game ends when the students offer your secret word as a test word. When they do, you can very dramatically reveal the word, showing them to be correct.

Issues That Arise in the Game

Be careful. An error in the count spoils the game very quickly. One technique is to say to one's self (silently), "Is the B in the test word? Is the R in the test word? Is the E in the test word?" Continue this way, counting the letters one by one.

A well-chosen secret word is important. Students quickly realize that a score of zero is very valuable. (Although they are at first surprised that a low score is actually better than a high score!) A score of zero eliminates letters and allows for cleaner tests of other letters. So you want them to get a score of zero quickly.

The most common letters in English words, in order, are ETAOINSHRDLU. For the first game it works well for the instructor to choose a word that includes the vowel U and some combination of the letters NSHRDL, together with letters that occur more rarely. Some good secret words are: CRUMB, BLUNT, CUBED and so on. It might be best to avoid the letter S, since many test words are plurals of four-letter words.

At some point it will become clear that an "elimination alphabet" is a good tool. Students write down an alphabet, cross out letters that are eliminated, and circle letters that are confirmed. If students don't think of this, you should at some point suggest it to them. It will serve to structure their thinking.

A B C D E F G H I J
K L M N O P Q R
S T U V W X Y Z

Classroom Discussion (during and after the game)

For the first few games, you can act as 'scribe', writing the results of the test words on the board. Get students to discuss the meaning of the test results: which letters are eliminated, which are definitely in the secret word, which letters need further testing, and so on.

What emerges is a notion of proof. Some students will "play a hunch," guessing that a certain letter is in the word. Others will give arguments that it must be, or cannot be, in the secret word. These are all good ways of getting students to think logically: to make deductions from known statements. This concept need not be made conscious. It is enough that students have the experience of engaging in logical deduction. A bonus for you is that they will be discussing their logic, and you can overhear and evaluate their stage of development.

Extensions

A student joins the teacher's team in thinking of a word and scoring:

Have a student come up and whisper a secret word to you. Then you play on the side of the student against the rest of the class. Check carefully that the test words are given the right score.

Students play in teams against each other:

Divide the class into two groups. Each must devise a test word, which they show you. You write it down and conceal it. They now play against each other, with you as moderator (checking their test scores!). They take turns offering test words. The team that guesses the other team's word first is the winner.

When students are playing against each other, each team should select a captain to mediate disputes as to what the next test word should be. They should also select a scribe, who takes over your duties in recording the test words and making notes on the elimination alphabet. But be sure to double-check the scores of the scribes. An error will ruin the game.

Errors in the Count

Such an error is often difficult to recover from. However, if it is caught early in the game (say after five or six test words are scored), you can go back, correct the scores, and proceed.

Sometimes it is clear that students are testing for a specific letter. In such a case, if an error has been made, actually telling students whether that letter is in or out of the secret word can avoid a logical tangle and allow the game to proceed.

But these are at best stopgap solutions. The game runs smoothly, and engages students best, if errors in the count are avoided.

Giotto for Experts

This game can be made more mathematical by considering each letter as a Boolean variable. Its value is 1 (present in the secret word) or 0 (absent in the secret word). One can then do "algebra" on the test words. In the example above (BREAD), we have BRINK = CRACK, with both words having a score of 2 . (Oddly, it is convenient here to use juxtaposition of letters to indicate addition, not multiplication.) But in fact the second C in CRACK doesn't contribute, so we have

BRINK = CRAK.

By cancellation of R's, we have

BINK = CAK.

That is, there are the same number of correct letters in BINK as in CAK, and so the score of either group is identical. We don't know if the score is 1 or 2 without seeing the secret word because the canceled letter, R, may or may not be in the secret word. (In this case, we know that R *is* in our secret word, so both BINK and CAK have a score of 1.)

Then we can cancel the K's, to get: BIN = CA.

Now we know that BIN and CA have the same score. If we were able, say, to eliminate CA (but in this case we can't) then we would know that the letters in BIN are also eliminated. We can use this information in a number of ways.

But this insight is only for a group that loves the game and gets very far into its analysis. Novices can enjoy and learn from the game without taking this advanced step.

The game can also be played with double letters allowed in the secret word. This version is considerably more difficult. For example, if the secret word is CREED, here are some sample test words and scores:

Test Word	
SHAME	1 (One E)
SHEER	3 (both E's and the R)
EERIE	3 (two of the three E's and the single R)

For most groups, the single-letter variant is more than enough for fun and learning.

Proof by Contradiction

Proof by contradiction is a quintessentially mathematical phenomenon. We begin by assuming that a statement is true. If we can deduce either a contradiction or a statement that is clearly false from such an assumption, then we know that we must have been wrong and that the original statement is in fact false. Students often use this sort of reasoning naturally and unconsciously, but it is useful to make it conscious as well.

A contradiction is a *relationship* among statements that must be false, no matter whether the statements themselves are true or false.

Students are sometimes uncomfortable with a statement that is a contradiction. They begin to think that they have themselves done something wrong. It is useful to show them that in fact they have been using the idea of a contradiction in solving other puzzles.

For example, you can help them recall the APE puzzle:

A + A + A + A + A + A + A + A + A + P = E

Here we immediately saw that A cannot equal zero, because then P would equal E. That is, we assumed that A = 0, and found that P = E. This contradicts the rules of our puzzle. We cannot have P = E (implied by our assumption) and also P not equal E (implied by the rules), so we have arrived at a contradiction. We can conclude that our original assumption is false, and A is not equal to zero. We then figured out that A = 1, etc.

We will see this kind of reasoning also occurring naturally in playing GIOTTO.

Indirect reasoning is often used to show that a certain situation cannot happen.

Example 1: Show that the cryptarithm BAA + BAA = EWE has no solution.

Solution: There must be a carry from the units to the tens column. To see this, we use indirect reasoning. If there were no carry, then in the units column, A + A = E, and in the tens column A + A = W. Then W would equal E, which is against the rules.

So there is a carry from the units to the tens column. Likewise there is a carry from the tens to the hundreds column. (In the units column we have A + A, and in the tens column we have A + A + 1 with the carry, so if there is a carry from the units column, there must also be a carry from the tens column.)

From all this we derive the contradiction that E is both even and odd. It is even because it is the last digit (or perhaps the only digit) of A + A. It is odd because there is a carry from the tens to the hundreds column, so E = B + B + 1.

Example 2: The four-digit number 4365 has an interesting property. If we write it backwards, and add the original number to the reversed number, we get a string of 9s: 4365 + 5634 = 9999.

a) Show that 2097 and 4185 each have the same property.
b) Find two more four-digit numbers that have this property.
c) Find a six-digit number that has this property.
d) Find a three-digit number that has this property.

The students and the instructor can work out the solutions to (a) through (c).

Solution to (d): There are no such three-digit numbers. Indeed, this problem is equivalent to solving the cryptarithm:

$$\begin{array}{r} A\ B\ C \\ +\ \ C\ B\ A \\ \hline 9\ 9\ 9 \end{array}$$

Let's assume first that there is a carry from the units digit to the tens digit. This leads to an impossibility, since we cannot get a sum of 19 by adding two digits (the largest sum we can get is 18). Let's assume the opposite, then, that there is no carry from the ones place. But then we reach another impossible situation. In the tens column we have B + B = 9, which is impossible since B + B must be an even number. We have shown that in either case we reach a conclusion that is obviously false and therefore that the puzzle has no solution.

Another way to see that there is no carry from the units column is to note that we are adding the same two digits in the hundreds column, and we don't have a thousands column, so there cannot be a carry.

Follow-up problem: For even *n*, show that there are many *n*-digit numbers with the property we are discussing. But for odd *n*, there are never any such *n*-digit numbers.

Lewis Carroll Puzzles

Lewis Carroll is well known as the author of Alice's Adventures in Wonderland, but he was also a mathematics professor with a particular interest in mathematical logic. The first puzzle below was taken from Lewis Carroll's diary (1894). Carroll created puzzles like these to train his students in logical reasoning. Like many wonderful math problems, they often seem to give too little information for a solution, but a unique answer emerges. Puzzles like these give students an appreciation of the power of logic.

A says B lies;
B says C lies.
C says A and B lie.

Who lies and who tells the truth?

Working this problem can lead to several interesting class discussions.

Reasoning from Assumptions

For students to do individually and in groups: Write the puzzle on the board. Have students try to (a) solve the puzzle and (b) convince other students that they are right.

As a whole class: Go over the puzzle together and have students share their thinking. An important step here is having students organize their thoughts and frame their arguments as precisely as possible.

Students sometimes don't know how to begin. We can facilitate their thinking by asking questions, such as, "Suppose we know that A is lying. What else do we then know?" Some students will respond that they know that B tells the truth (from statement 1). That means that C is indeed lying (from statement 2). But then either A or B must be telling the truth. We've assumed it's not A, so it must be B. So if A lies, B tells the truth and C lies, and we're OK.

In fact, the key to this solution is the assumption that A is lying. This leads to a simple solution, but it is not easy to find this starting point.

It is important to note that this is not the only way to solve the problem. Students starting with some other assumption will eventually reach the same conclusion. The idea that another argument will lead to the same conclusion is a good starter for conversation.

This is one possible solution to the puzzle, and students are typically satisfied with it. But is it the only possible solution? Let students enjoy their solution before raising this embarrassing question, which will be settled later in a more formal approach to the problem.

Many alternative solutions use the principle of proof by contradiction. Proof by contradiction can be difficult for students to understand. If some students don't understand, just proceed with those who do, and cycle back later to the notion. It will come up naturally in the solution of many of the puzzles here.

As an example of how proof by contradiction comes up naturally in this puzzle, suppose we start with the assumption that A is telling the truth. Then we know that B is lying (from statement 1). Now B asserts that C lies (statement 2), so C must be telling the truth. If C tells the truth, then A lies (from statement 3). We have reached a contradiction: Starting with the assumption that A is telling the truth, we conclude that A lies. Thus our assumption is incorrect, and A must be lying. The rest of the solution proceeds as above.

Here's a slightly more complicated example of indirect reasoning that might emerge:

Let's assume that C is telling the truth. Then A and B must both lie (from statement 3). But if A lies, then this means that B is telling the truth (from statement 1). And if B is telling the truth, then C is lying (from statement 2). We started by assuming that C tells the truth, and end up deducing that C lies. This is a contradiction, so C cannot be telling the truth; i.e., C lies. But if C lies, then B is telling the truth because (from statement 2) B says C lies. So B must be telling the truth, and (from statement 1) A is lying. With different reasoning we have again reached the conclusion that A lies, B tells the truth, and C lies.

Similar reasoning holds whether we assume that B lies, or B tells the truth, or whether we assume that C lies, or C tells the truth.

There is no need to go over all of this explicitly in class. We mention these possibilities because the different arguments may come up in students' reasoning.

Structuring the Reasoning

One interesting property of this sort of logic is that it is highly susceptible to automation. That is, when we structure the thinking, we can often find ways to make the process mechanical. This eventually leads to the phenomenon of "thinking machines": computers.

It is important not to proceed too quickly to this next stage in discussing the problem. It is useful for students to have a fluid and thoughtful conversation before the problem is made mechanical. The intuitions need to be developed before they are captured in formality.

Eventually, some students will find a way of writing down their reasoning. If students don't come up with one, show them the following way of structuring and recording their analysis. We have used T to mean the person is telling the truth and L to mean the person is lying. DO NOT show students how to fill in the table. That part is next.

	A	B	C	Possible?
1.	T	T	T	N
2.	T	T	L	No
3.	T	L	T	No
4.	T	L	L	No
5.	L	T	T	No
6.	L	T	L	YES!
7.	L	L	T	No
8.	L	L	L	No

1. Students work individually to construct the blank the table.

It is a worthwhile activity in itself for students to list all 8 possibilities. (Because there are 3 people and there are 2 possibilities for each, there are 2 × 2 × 2 total possibilities.)

It is useful to have a "canonical" way of listing the possibilities. That is, once students understand that there are eight possibilities for truth-tellers and liars, suggest to them that they be listed in the order given above (or any other order you choose, so long as it is the same for all students).

2. Students use proof by contradiction to rule out possibilities and fill in the chart.

Have students consider each possibility and see whether it results in a contradiction. As they reason, they can fit their thinking into the chart to verify that there is only one possible answer.

Listing all the possibilities like this makes things much clearer, and shortens many arguments. For example:

Assume that B is lying.

B lies -- > C tells the truth -- > A lies -- > B tells the truth (Contradiction!)

Now we can cross off *any* possibility on the table in which B is lying, and so on.

3. Facilitate a class discussion to make students aware of their use of "if...then" reasoning and "proof by contradiction."

Students will find themselves arguing, for example, "Suppose A is telling the truth – then what? That can't be because it leads to a contradiction . . . ."

Next:

If exactly one of the following is true, which is it?

(a) Puppies are cuter than kittens.
(b) Puppies are cuter than bunnies and kittens.
(c) Puppies are by far the cutest animals.

If students are stuck, give them the hint that the key phrase is "exactly one."

From there we use a proof by contradiction.

If we assume (c) to be true, then (b) and (a) are true too. That contradicts the statement that only one is true.

If we assume (b) to be true, then (a) is true too. That also contradicts the statement that only one is true.

Therefore (a) is the only one that can be true. It is the only one that doesn't lead to a contradiction.

This conclusion is enough for the problem to be worthwhile. But if a class is interested, we might go further. If (a) is the only statement that is true, then (c) must be false: there must be some animals cuter than puppies. And (b) must also be false, which means that bunnies are cuter than puppies. So we can infer a situation about animals, not just about statements: Bunnies are cuter than puppies, which are cuter than kittens.

Of course all this follows from the premise that exactly one of these statements is true, a premise that may have nothing to do with reality. Mathematical truth is logical, and not empirical.

If exactly one of the following is true, which is it?

(a) 10 year olds are delightful.
(b) 11 year olds are delightful.
(c) Either 10 year olds or 11 year olds are delightful.
(d) Blue is the loveliest color.

Solution: Again we use proof by contradiction.

If (a) is true, then (c) is also true, contradicting the statement that exactly one of the statements is true.

The same argument holds if (b) is true.

If (c) is true, then we know that at least one of (a) or (b) is true. Although we don't know which, we get a contradiction anyway: more than one of the statements would have to be true.

Hence only (d) can be true.

Building the Meaning of Implication

The fundamental task of any young student of logic is to get a handle on the notion of *implication*, of what it means for statement A to *imply* statement B. This is not so simple, and we are just looking at the easiest examples here.

We can use the language of implication by formulating the following principle with the students:

If statement A implies statement B, then every time A is true, B must also be true. So in a list of statements, of which only one can be true, any statement in the list that *implies* another statement in the list must be false.

"Application" to Standardized Tests

This principle of logic can be applied to the "students' reality" of the multiple-choice test. In a well-constructed test of that form, there is exactly one right answer. That is, the list of answers is a list of statements, exactly one of which is true.

So, by the principle above, any choice that implies another choice must be wrong.

Bernard is thinking of a number.
(It is a whole number.)
Which one of the following is true:

(a) The number is greater than 7
(b) The number is greater than 8
(c) The number is greater than 12
(d) The number is greater than 15.

The solution hinges on the logical principle we outlined above. Statement (d) implies statement (c): if a number is greater than 15, it is certainly greater than 12. Similarly, (c) implies (b), and (b) implies (a). Thus (a) must be the correct answer.

Notice that we can tell something else. We know Bernard's number!

Once we conclude that the number is a whole number greater than 7 but not greater than 8, we know it must in fact be 8.

Have students discuss this unusual phenomenon. It occurs not because we can read minds, but because Bernard is a character in a multiple-choice test.

Class Discussion

After students have worked through these puzzles, bring them together for a class discussion. What does this show about logic?

The main ideas here are:

We can use proof by contradiction. Suppose a particular statement implies a statement that must be false (because we know that some third statement is true). Then the statement we started with must in fact be false.

The essence of logical deduction is an "if-then" statement. (If A is true, then B must also be true.) We cannot expect a full formal understanding of this concept right now, but we can build towards it, by using the word "implication" correctly to describe the relationship between two statements.

Logic is not about truth or falsehood of specific statements, but about relationships among statements. For this reason, logic can help us to derive new information from previous statements, and sometimes the results are surprising.

More Puzzles from Lewis Carroll

The puzzles below were included by Lewis Carroll in his book, Symbolic Logic (1897). Carroll wrote the puzzles for university students of formal logic, and he created his own system of symbols and diagrams for analyzing logical relationships such as these. We have included a few of the simpler examples here, which are accessible to young people and do not require the use of formal symbols. Students may find it useful to use Venn diagrams to analyze the problems, or they may invent their own symbols and diagrams. Notice that Carroll's puzzles include statements that are silly, empirically unverifiable, or even false. The puzzles are amusing, of course, but Carroll also wanted to be sure that his students would solve the puzzles with pure logical reasoning and not with reference to any background knowledge.

Worked example:

(a) All babies are illogical.

(b) Nobody is despised who can manage a crocodile.

(c) Illogical persons are despised.

What can we conclude?

____

Solution: Each of the statements above can be represented with a Venn diagram:

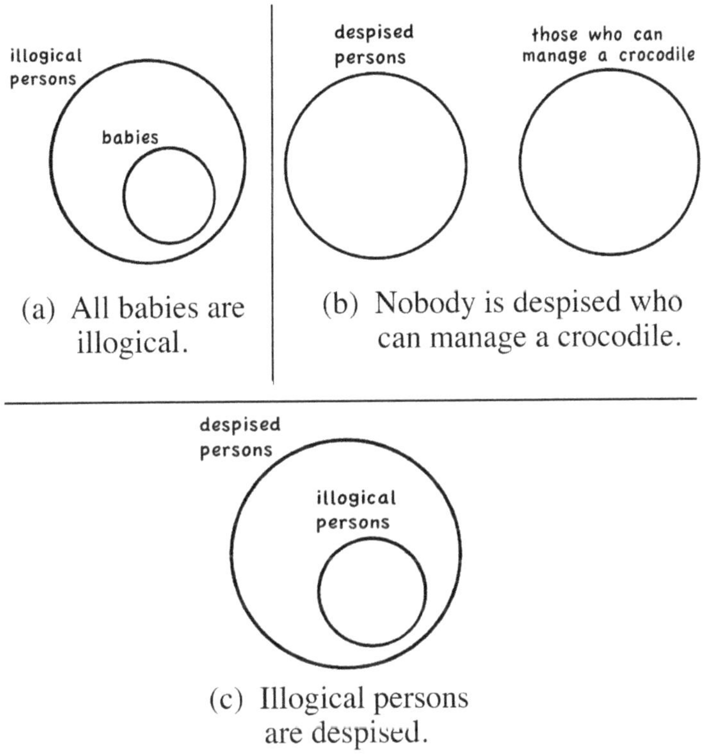

(a) All babies are illogical.

(b) Nobody is despised who can manage a crocodile.

(c) Illogical persons are despised.

And we can combine them:

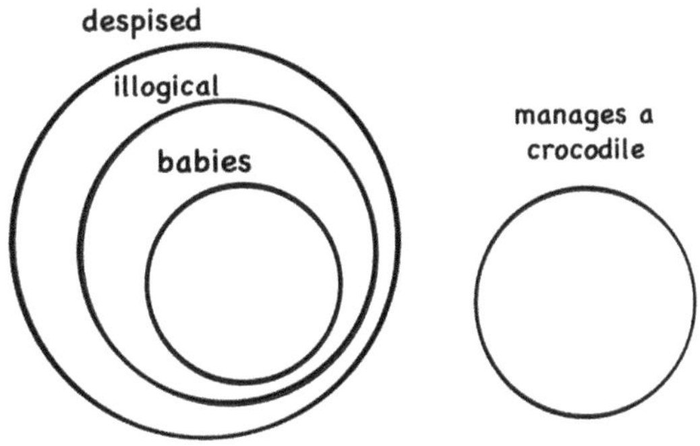

Notice that, among other things, we can conclude:

There is no baby that manages a crocodile,

no illogical persons manage a crocodile, and

all babies are despised.

Your Turn!

1. (a) No ducks waltz.
   (b) No officers ever decline to waltz.
   (c) All my poultry are ducks.

2. (a) All puddings are nice.
   (b) This dish is a pudding.
   (c) No nice things are wholesome.

3. (a) My saucepans are the only things I have that are made of tin.
   (b) I find all *your* presents very useful.
   (c) None of my saucepans are of the slightest use.

4. (a) No potatoes of mine, that are new, have been boiled.
   (b) All my potatoes in this dish are fit to eat.
   (c) No unboiled potatoes of mine are fit to eat.

5. (a) No one takes in the *Times*, unless he is well educated.
   (b) No hedgehogs can read.
   (c) Those who cannot read are not well educated.

6. (a) No birds, except ostriches, are 9 feet high.
   (b) There are no birds in this aviary that belong to anyone but *me*.
   (c) No ostrich lives on mince pies.
   (d) I have no birds less than 9 feet high.

Solutions

1. (a) No ducks waltz.

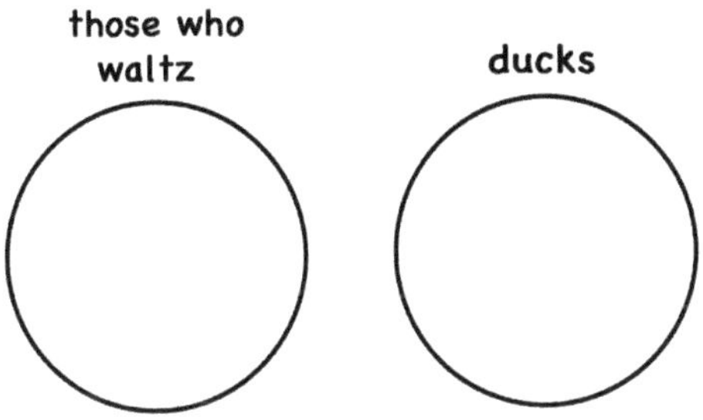

(b) No officers ever decline to waltz.

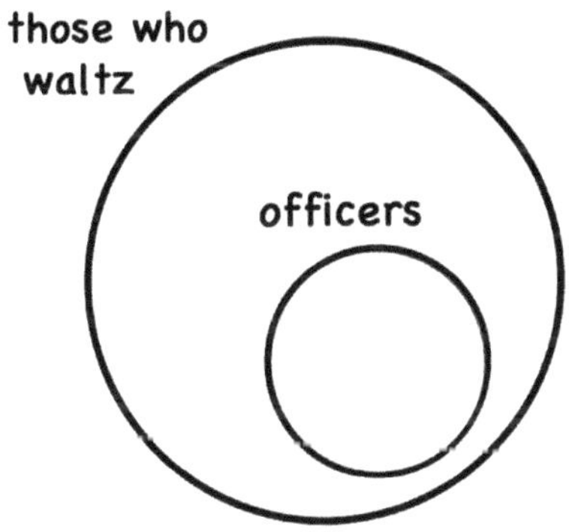

(c) All my poultry are ducks.

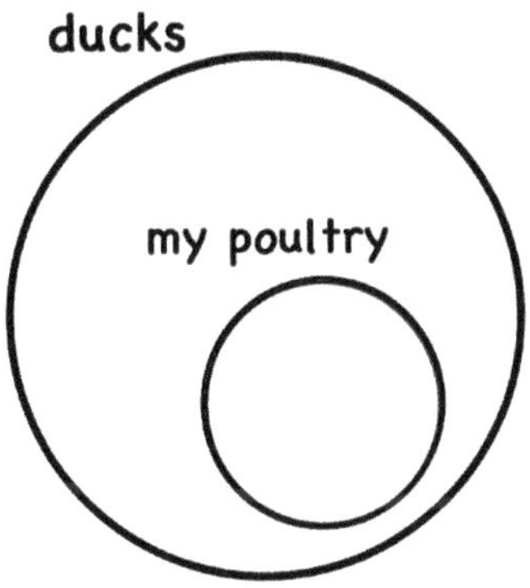

All together:

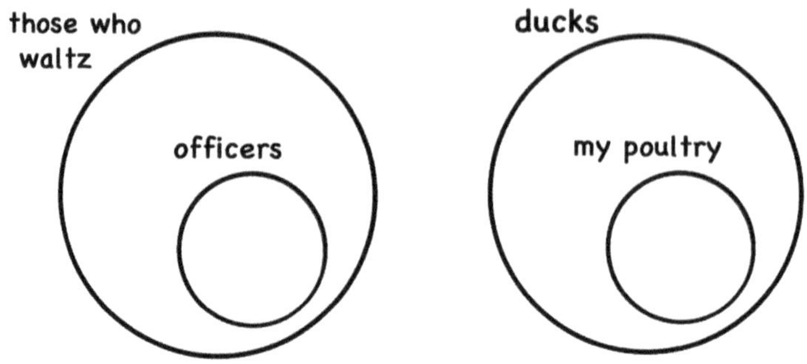

We can conclude, among other things, that my poultry are not officers and that they don't waltz. Also, we can state with certainty that no officer is a duck.

2. (a) All puddings are nice.
(b) This dish is a pudding.
(c) No nice things are wholesome.

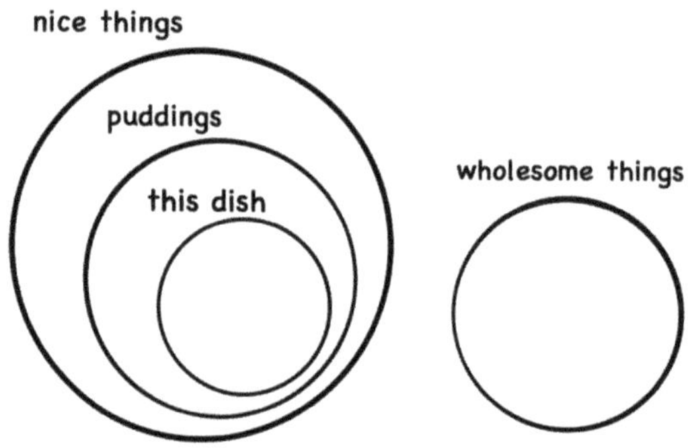

We can conclude that puddings are not wholesome and also that this dish is not wholesome.

3. (a) My saucepans are the only things I have that are made of tin.
   (b) I find all *your* presents very useful.
   (c) None of my saucepans are of the slightest use.

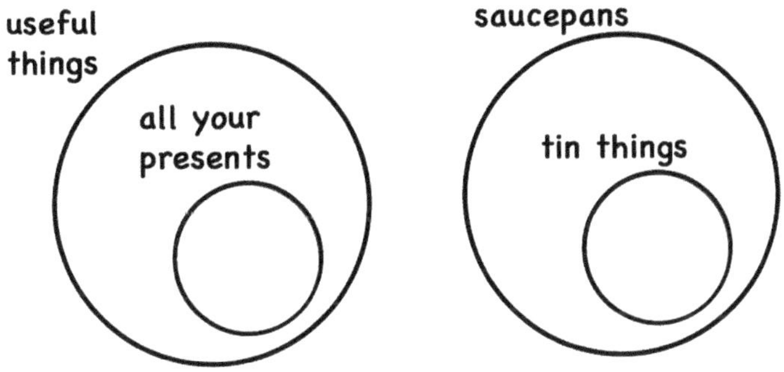

We can conclude, for example, that you never gave me anything made of tin.

4. (a) No potatoes of mine, that are new, have been boiled.
   (b) All my potatoes in this dish are fit to eat.
   (c) No unboiled potatoes of mine are fit to eat.

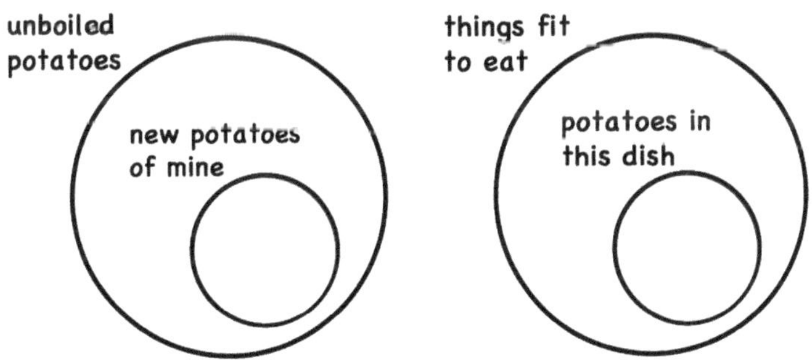

We can conclude, for example, that the potatoes in this dish have been boiled, and they are not new. We can also conclude that my new potatoes are not fit to eat.

5. (a) No one takes in the *Times*, unless he is well educated.
   (b) No hedgehogs can read.
   (c) Those who cannot read are not well educated.

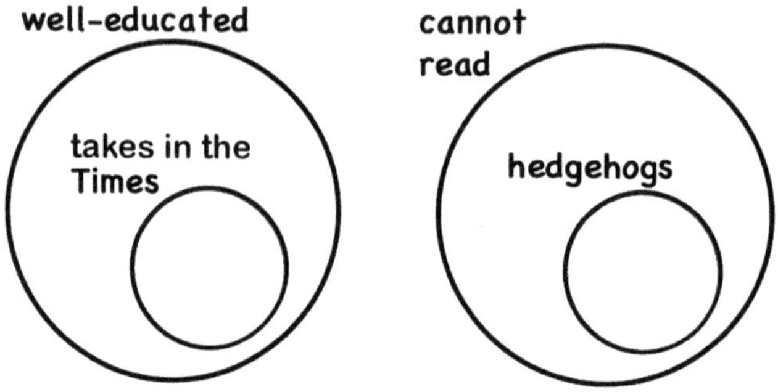

We can conclude, for example, that hedgehogs are not well educated and do not take in the *Times*. (Students may not understand the meaning of "take in the *Times*." It may be best to have students solve the puzzle even before explaining the meaning of the phrase. This helps emphasize the point that the logical deductions themselves do not depend on the truth of the content of the underlying statements.)

6. (a) No birds, except ostriches, are 9 feet high.
   (b) There are no birds in this aviary that belong to anyone but me.
   (c) No ostrich lives on mince pies.
   (d) I have no birds less than 9 feet high.

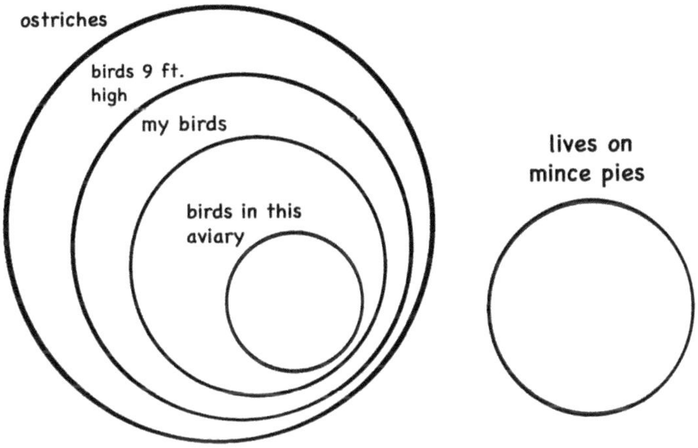

We can conclude, for example, that no birds in this aviary live on mince pies.

Cryptogram Puzzle

In this puzzle each letter stands for another letter. Look for patterns in the words to figure out what the letters represent. For example, vowels occur more often than consonants. The location of commas and apostrophes can also give us clues. Students may also notice that the letters "iy" appear together a lot, and they can think about letter pairs that often go together. (In this case the letter pair represents "th.") The greatest tool we have here, aside from looking for these sorts of patterns, is "proof by contradiction." If something seems reasonable, then assume it to be true and see if it leads to an impossibility.

One easy place to begin the solution is with the single letter B in the first panel. The letter stands by itself, and there are only two one-letter words in English: "a" and "I." If we assume that B stands for "a," we can easily reach a contradiction. For example, in the second panel, we have the code word B'N. But there is no contraction in English starting with the letter "a." So B cannot be "a" and must stand for "I."

There are numerous paths to solution, all involving making use of logical deduction—together with the students' knowledge of English. (We have found that even students who are not native speakers of English can work this puzzle.)

Note: Normally we would see the letter "u" by itself and assume it must be an "a" or an "I." Here, it is "*x*," which could lead students astray. It may be a good idea for the instructor to talk this out with the class and help them see that their assumption may not be true in this case. The letter "u" appears only one other place in the puzzle – at the end of a word. It must be a more unusual letter.

Solution:

Puzzle letter	Actual letter
B	I
C	R
D	A
E	J
F	S
G	B
H	K
I	T
L	U
M	D
N	M
P	E
Q	N
R	W
S	F
T	O
U	X
X	Y
Y	H

Solution:

Mrs. Smith, I know the answer! "x" is three!

No, I'm afraid the answer is fifty-six.

But Mrs. Smith, just yesterday you said "x" was three.

Day 2

Goals for the Day

Students will continue practicing their skills of logical deduction, and will be introduced to two new mathematical principles: invariants and parity.

Giotto, Part 2

The following exercises ("Giotto puzzles") make explicit the reasoning students used the previous day in playing the game of Giotto. Students can solve these puzzles alone or in pairs and then discuss as a group.

In a game of Giotto, what can you conclude, if anything, from the following scores? Each problem is a separate puzzle.

Problem #1:
PLANS 1
PLANE 1

Solution: We can't tell much. There are two possibilities: (1) The letter in common with the secret word is P, L, A, or N; or (2) the secret word has both an S and an E. We cannot rule out any particular letter. We can see this by testing each letter and looking for a contradiction. Test by saying, "Could there be a P in the secret word? Could there be an L? An S? An E?" None of these possibilities leads to a contradiction.

Problem #2:
SLAPS 2
LOVES 0

Solution: Here we can tell a lot. We know the mystery word has no L, O, V, E, or S. And, looking at SLAPS, if there is no S and no L, there must be an A and a P.

Note the importance of a score of 0.

Problem #3:
FIGHT 2
NIGHT 1

Solution: Here we can tell a lot, too. We can use a proof by contradiction to show that there is no N. First, let's assume there is an N in our mystery word. Then there would be no I, G, H, or T. If there were no I, G, H, or T, then FIGHT would have at most 1 point. But that leads to a contradiction because FIGHT shows 2 points.

We can also tell that because the mystery word has no N, it must have exactly one of the following letters: I, G, H, or T. And if it has only one point from I, G, H, or T, then it must also have an F.

For experts: The "algebraic" method shows this mechanically. Considering each letter as a variable with value 0 (absent) or 1 (present), we have FIGHT = NIGHT + 1. (Recall that here the juxtaposition of letters means addition and not multiplication.) Then we can subtract IGHT from each side, to get F = N + 1. If N = 1 (if N were present), then F would equal 2. But these variables cannot take the value 2. This contradiction shows that N cannot be present.

Further, the "algebra" tells us that F = N + 1 = 0 + 1, so F is present in the word, and N is absent.

Problem #4:
WORDS 1
WARDS 3

Solution: This set of scores is not possible. If the mystery word has three of the letters in WARDS, then at least two of those letters must also be in WORDS. But that leads to a contradiction because WORDS shows only 1 point.

In algebra, WORDS + 2 = WARDS, or O + 2 = A. But A is at most 1, so this cannot happen.

Problem #5:

PLANT	1
PLANK	0
FORTE	1
THREE	2

Solution: Here we can tell quite a bit.

If there is no letter in PLANK, then the point showing for PLANT must come from the T.

If the mystery word has a T, then it also has no F, O, R, or E because FORTE shows only 1 point.

And if it has no R or E, then the 2 points in THREE must come from the T and the H.

So we know two mystery letters: T and H.

The Language of the Watermelons

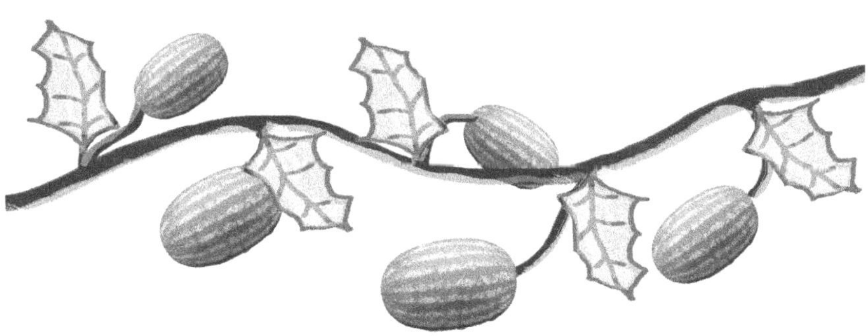

Teaching Notes

Give students the Watermelon worksheet and introduce the activity. Students can work mostly on their own while the instructor brings them back together from time to time to discuss what they have figured out and to make sure everyone is up to speed.

The activity is an application of logic to number systems and to linguistics. Students are asked to reason about the structure of a number system and a language based on clues they are given.

For Students (Part I)

Watermelon Language

Watermelons talk to each other in the garden, when no one is listening. They have their own language, and my friend Tubby Green (a watermelon) taught me some of it. But only a little.
He's not really good at mathematics. Here's what he told me:

1) Even numbers end in *–ina.* Odd numbers end in *–iddle.*

2) The numbers 1, 2, 3, 4, and 5 are:

Diddle
Fiddle
Riddle
Amina
Semolina

But of course not in that order.

(Can you give a simple reason why the number names above cannot be in order?)

3) Diddle + Diddle = Amina

4) Diddle + Semolina = Fiddle

a. Can you figure out which number is which, in the watermelon language?

For (b) through (g), give your answers in the watermelon language.

b) What is Diddle + Riddle?
c) What is Amina × Amina?
d) What is Fiddle – Riddle?
e) Tubby said that Riddle + Riddle = Middle. But I think he's wrong. How can I tell, from the rules given above?
f) Tubby said that Amina + Semolina = Ocarina. Can he be right?
g) If Tubby is right in (f), what is half of Ocarina?
h) If Tubby is right in (f), what is Ocarina – Fiddle?

Solutions to Part I

a) The even numbers are Amina and Semolina because they end with *-ina.* But which is which?

We know that Diddle is odd, and twice it is less than 5. So Diddle must be 1. This implies that Amina = 2. Then Semolina (the other even number) = 4, and Fiddle = 5. The value 3 is left to Riddle:

Diddle	1
Amina	2
Riddle	3
Semolina	4
Fiddle	5

b) Diddle + Riddle = 1 + 3 = 4 = Semolina.

c) Amina × Amina = 2 × 2 = 4 = Semolina.

d) Fiddle – Riddle = 5 – 3 = 2 = Amina.

e) Tubby cannot be right. Riddle + Riddle must be an even number, and so must end in *-ina.*

f) Amina + Semolina = 2 + 4 = 6. Tubby can be correct if Ocarina is 6.

g) Half of Ocarina is 3 = Riddle.

h) Ocarina – Fiddle = 6 – 5 = 1 = Diddle.

For Students (Part II)

Tubby told me a bit more about Watermelon Language. (He was right about Ocarina by the way: it's twice Riddle.) Tubby said:

Ocarina + Diddle = Ocadiddle
Ocarina + Riddle = Ocariddle
Amina × Semolina = Ocamina

Can you count to 11 in Watermelon Language?

Solution to Part II

Ocarina + Diddle = Ocadiddle 6 + 1 = 7 = 6 more than 1
Ocarina + Riddle = Ocariddle 6 + 3 = 9 = 6 more than 3
Amina × Semolina = Ocamina 2 × 4 = 8 = 6 more than 2

So the prefix *oca-* means "add 6."

Can you count to 11 in Watermelon Language?

1 Diddle
2 Amina
3 Riddle
4 Semolina
5 Fiddle
6 Ocarina
7 Ocadiddle
8 Ocamina
9 Ocariddle
10 Ocasemolina
11 Ocafiddle

...so it's a base-6 system.

Extensions

Have students extend the number system.

Notice that in our base-10 system, *-teen* means "add 10," as in:

Fifteen = Fif-teen = 5 + 10
Sixteen = Six-teen = 6 + 10
Seventeen = Seven-teen = 7 + 10,

and so on.

And *-ty* means "times 10," as in:

Fifty = Fif-ty = 5 × 10
Sixty = Six-ty = 6 × 10
Seventy = Seven-ty = 7 × 10,

and so on.

For a base-6 system, we will need a new suffix (or prefix) to mean "times 6."

Here, we will use the suffix *-ocky* as an example. But students can make up their own words.

12 = Amina-ocky = Aminaocky (or even Aminocky)
(which means two times 6)

13 = Aminocky Diddle (which means two times 6 plus one)
(Or, we can include a hyphen: Aminocky-diddle. This would be analogous to "twenty-three."

14 = Aminocky Amina (which means two times 6 plus two)

15 = Aminocky Riddle (which means two times 6 plus three)

16 = Aminocky Semolina (which means two times 6 plus four)

17 = Aminocky Fiddle (which means two times 6 plus five)

18 = Riddleocky (which means three times 6)

etc.

And what about 36? We would need a new word for 36 (six 6's), just as we need a new word for 100 (ten 10's) in a base-10 system.

Groups of students sometimes enjoy continuing this sort of activity in another base. They love making up a system that is their own.

The Jittery Soldiers Problem

This classic problem is simple to state but introduces some profound ideas. We will use it to introduce the idea of an *invariant*, a logical principle that is used often in mathematics. This problem can easily fill a few hours if it is fully explored.

Introducing the problem to the group by acting it out

Instructor explains the problem and then brings five students to the front of the room to act it out. Together the group talks through a few examples.

<u>The Problem</u>: Five soldiers stand in a line, while the sergeant yells orders at them. The sergeant yells, "Left FACE!" The soldiers are a bit jittery (from being yelled at) and a bit confused, so they don't always follow the order. Some turn to their left and others turn to their right.

Each soldier can only see the soldier directly ahead of him after he executes the order.

Example 1: The sergeant yells, "Left FACE!" Soldiers 2, 3, 4 and 5 turn to their left. Soldier 1 gets it wrong and turns instead to his right.

(In the illustrations it is assumed that the soldiers begin facing the reader who is in the position of the sergeant, and that the "R" and "L" refer to the soldiers' right and left.)

<u>Question for the class</u>: Do any of the soldiers realize that there is a problem?

No. Each soldier notices only the soldier next to him, in the direction he is facing. So for this example, each left-facing soldier sees another soldier facing in the same direction, and the sole right-facing soldier sees no one at all. None of them sees that someone else is facing a different way.

Example 2: The sergeant yells, "Left FACE!" This time soldiers 2, 3, 4, and 5 get it wrong. Only soldier 5 is correct.

Question for the class: Do any of the soldiers realize that there is a problem?

This time, soldiers 1 and 2 realize that there is a problem. Since they are facing each other, one of them must have done the wrong thing.

Example 3: The sergeant yells, "Left FACE!" Soldiers 1, 3 and 5 turn correctly. Soldiers 2 and 4 turn to the right instead of to the left.

Question for the class: Which soldiers notice a problem?

Soldiers 1 and 2 are facing each other, as are soldiers 3 and 4. So each of these pairs knows something is wrong. Soldier 5 (who happens to have been correct) doesn't notice anything at all.

It is not hard to see that the only soldiers who notice a problem are those who might be standing face-to-face.

Now we add the condition that if two soldiers are standing face-to-face, they suddenly realize that an error has been made. They *each* do an about face, exactly one second later (because neither knows which of them has made the error!).

Let's follow the situation in example 3 to see what ensues. The places where soldiers are looking at each other and notice a problem are shown.

After 1 sec.	L	R	L	R	L
after 2 sec.	R	L	R	L	L
after 3 sec.	R	R	L	L	L
	1	2	3	4	5

...and after three seconds the situation remains stable: none of the soldiers notices anything wrong.

There are several questions that can now be explored. For example:

a) What do the stable situations have in common?

b) Can we predict from the initial situation what the stable situation will look like?

c) How long will it take to reach a stable situation?

And there are many others. We will address (a) and (b) through the problems below.

Acting out the problem in class helps students understand it. (And it is fun.) Have a group of five students arrange themselves at random. The teacher can then explain which "soldiers" notice a problem, and how they move. For practice the teacher can call out, "Left FACE" initially, then call out, "I said, 'Left FACE!'" at regular intervals (probably longer than one second!) as the students rearrange themselves. The class can check that the procedure is being carried out correctly and note when a stable position has been reached.

Students choose their notation and solve the problem on paper

Next, students will find their own way to represent the problem symbolically, and then will work out on paper the problem they just acted out.

Up until now we have acted this out with students in the front of the room. Now ask students to think of a way to represent the problem on paper. We have been using L's and R's, but any other pair of symbols will do. Have students make suggestions and agree as a class upon which symbols to use. (A typical symbol-pair might be 0 and 1, < and >, etc.) While there are always many ways to represent a problem, some are better than others. Here, for example, students may want to draw the soldiers for each iteration of the problem. While this might help initially to see what's going on, for further work it becomes inefficient.

For the rest of the exercise, we will use 0 for soldiers facing their right (the reader's left) and 1 for soldiers facing left (the reader's right).

Question 1: Use paper and pencil now to work out one of the examples we acted out in the front of the room: What happens if we start with 10000?

Solution:
10000
01000
00100
00010
00001, which is stable.

Question 2: Work a few more examples on paper:

Here are a few examples with solutions:

11100 initial condition
11010
10101
01011
00111 stable

11010 initial condition
10101
01011
00111 stable

01000 initial condition
00100
00010
00001 stable

Notice that every time we see "10" in the line, those two soldiers reverse positions.

Question 3: We have now seen some different stable arrangements, including 00001 and 00111. Can you find others? How many stable arrangements are there all together? What if there were six soldiers? One hundred soldiers?

Solution: For five soldiers, the stable conditions are:

00000
00001
00011
00111
01111
11111

Notice that the stable positions have some number of 0's on the left, then some number of 1's on the right. The 0's and 1's are "sorted."

Question 4: (Extra credit) For five soldiers, how many total positions are possible (stable or unstable)?

Solution: 2 x 2 x 2 x 2 x 2 = 32

There are many ways to approach this combinatoric problem. Students who have seen such problems will use one of the methods they know. But if students haven't had any experience with combinatorics, this question may be too much of a diversion. Its solution is not used in later work.

Question 5: Look at the sequences you created. The soldiers' directions change in each line, but what stays the same? (In mathematics, this is called an "invariant.")

(This question is pivotal for student learning from this activity.)

Solution: In each sequence, any 10 switches and becomes a 01. So, while the direction changes, the number of 1's and 0's stays the same. The number of 1's is an invariant.

(There are, of course, other invariants: the number of soldiers, the number of neighbors each soldier has [one or two], and so on. Try to elicit from the class this invariant, which is the most meaningful one.)

Question 6: How can we predict based on the starting positions how things will end up?

Solution: The only stable positions are those with all 0's on the left and all 1's on the right. Because the number of each is invariant, the ending position must have the same number of 0's as the starting position, but bunched up on the left. The rest of the positions are 1's.

Question 7: What if we start with a line of one hundred recruits, and all the odds turn right and the evens turn left? How would they end up?

Solution: There are fifty of each. We would end up with fifty 0's (on the left) and then fifty 1's (on the right). Notice that if we had started with that big problem, a solution would have been extraordinarily difficult. An understanding of the simpler version of the problem makes the harder problem easy. This is a good problem-solving trick: make a smaller "model" problem, which may help solve the larger problem.

Class Discussion

What kinds of logical arguments are we using in this problem? One is the idea of an "invariant," an idea that is often used in mathematical proofs. An invariant is some aspect of a mathematical object that remains unchanged as we apply changes to the object.

A simple example:

Suppose we take a rectangle and stretch it in a direction parallel to one pair of edges. What changes and what remains the same? The rectangle still has four sides, and the angles are still right angles. Those are invariants under this kind of change. On the other hand, length changes, area changes, and perimeter changes.

If we are allowed to tilt the sides so that the angles change, this invariant (angle measurement) is dropped. But the lengths of the sides remain the same. More subtly, the sides remain parallel. The figure is still a parallelogram.

On the other hand, if we think of the sides as made of rubber bands and can stretch them without breaking, crossing, or touching (in the plane), then the rectangle can be deformed into a circle, an oval, or even a pentagon or triangle. But the figure still divides the plane into two regions (an outside and an inside). The new figure will not, for example, become a figure 8 (which divides the plane into three regions). The number of regions into which the plane is divided is invariant.

These are all examples from geometry. In fact, one way to describe the study of geometry is by specifying a set of allowable transformations and studying the invariants of this set of transformations. That is, of course, a much deeper study than we have here. But the Jittery Soldiers can start students thinking about a very deep concept in mathematics.

The Black-and-Red Card Problem

This card problem is an example of a situation in which an invariant is useful. It is also a good introduction to parity, another useful idea in mathematical proofs. At first the students will just play the "game" in pairs, get used to the rules, and observe the outcomes. The next day the instructor will give them more explicit instruction about parity and how it applies to this problem.

The rules:

(1) Begin with three black cards and four red cards.

(2) Turn over two cards at a time and set them aside.

(3) If you turned over a red card and a black card, then set them aside and replace them with a single black card.

(4) If you turned over two cards of the same color, then set them aside and replace them with a single red card.

The rules can be represented simply in a chart:

If we remove,	Then we add:
RB	B
BR	B
BB	R
RR	R

The problem: What will the last card be? In play, students will observe that the last card is always black.

Next, change the game by starting with three black cards and eight red cards, and ask students to predict the outcome. Again, students will observe that the last card is always black. (This is a surprising result because there are many more red cards.)

The analysis should be done the next day, after a lesson in parity.

Day 3

Goals for the Day

Students will gain more experience with different applications of the concept of "parity." Students will also be introduced to the pigeonhole principle and will see powerful applications of the principle. They will also see that we can gain insight into a difficult problem by looking for a simpler, isomorphic problem in the Mouse and Cheese Problem.

Parity

Parity refers to the evenness or oddness of an integer. For example, the parity of 4 is even, and the parity of 193 is odd.

For more advanced students, we can point out that we need not know a number to know its parity. A description of the number may suffice. The parity of the number of married people in the world is even, if we assume that no one is married to more than one other person at a time. The parity of the number of shoes produced in a particular factory is (probably) even: it is not likely that a left shoe is produced without a right shoe being produced. The parity of a committee of judges should be odd, so that there cannot be ties in the judging. Students can make up other examples in class discussion.

For younger students, spend more time talking about evens/odds to be sure they understand. It is important for them to understand that zero is an even number because it comes up in some of the proofs. There are different ways to see that zero is even. Students sometimes observe the alternating pattern even, odd, even, odd and extend it to conclude that zero must be even. Sometimes a student will argue that zero is even because it is a "double number": $0 \times 2 = 0$.

Parity is often used in proofs to show that something is impossible. Sometimes, for example, we can tell that the parity of two numbers is different even if we don't know what the numbers are. If so, we know that those two numbers can never be equal.

The Five Cups Problem

This is a classic mathematical puzzle. Begin with five cups placed like this:

Four cups are placed upright and one is placed upside down. A move consists of turning over two cups at a time. The goal is to turn all of the cups upright.

Teaching notes: Give students plenty of time to understand the rules and then to work through the puzzle. Students may see that it is impossible, but the instructor should encourage them to think more deeply and try to articulate why it is impossible. Students can be given the hint that the class is studying *parity* and therefore that they should try to think in terms of odds and evens.

Solution: The number of cups initially turned up is even. Because the parity never changes as we make moves, we can never have an odd number of upturned cups.

Why doesn't the parity change? No matter what move we make, we are either adding two to the number of upturned cups (even + 2 = even), taking two away from the number of upturned cups (even – 2 = even), or leaving the number unchanged (even + 0 = even).

That is, the parity of the number of upturned cups is *invariant*.

Extension #1: Is the problem solvable if we start with five cups turned up and one turned down? Students will say initially that it is possible because now we have an even number of cups. However, we now have an *odd* number of upturned cups, and again the parity never changes. The two quantities (the number of upturned cups and the total number of cups) can never be equal.

Extension #2: Start with six cups. Is the problem solvable with two turned down? With three turned down? And so on. In general, if you have an even number turned down, then the parity of the upturned cups matches the total number of cups, and there *is* a solution.

Extension #3: Can we solve the puzzle if we are allowed to move three cups at a time? Yes. The parity is not constant this time, and we can solve the puzzle:

DDDDU

DDUUD

UUUUU

A subtle point arises here, which can be explored if students are ready. We can see that parity does not remain invariant if we turn three cups at a time. But that does not mean "automatically" that the puzzle can be solved. In *this case,* we can solve it. In fact, we can prove that any arrangement of cups can be turned into any other arrangement if we are allowed to turn three cups at a time. (The proof is not so easy for this level of student.) But in a different situation, there might be some other circumstance preventing us from solving the puzzle, even if we can change the parity. For example, there might be some other invariant that prevents a solution.

The Twenty-Five Flowers Problem

Henry has twenty-five flowers and wants to place them into four vases. He can place any number of flowers in each vase, but the vases must either all have an odd number of flowers or all have an even number of flowers. How can he do it?

Solution: The sum of either four even numbers or four odd numbers will be even. Therefore, it is impossible to place twenty-five flowers in the vases in this way. In fact, we can generalize by saying that this will be true for any even number of vases and odd number of flowers.

Extension: Is there a solution if we have an odd number of vases and an even number of flowers?

Solution: In this case the parity of the numbers does not prevent a solution, but we have to be careful about how we distribute the flowers among the vases. If we place an odd number of flowers in each vase, the sum will be odd. So the only solutions will have an even number of flowers in each vase.

The Covetous Neighbors

This problem was one of the weekly puzzles from NPR's Car Talk.

Suppose that there are twenty-five neighbors living in separate apartments in a building that is the shape of a five-by-five grid, as in the diagram above. Each person is jealous of his adjacent neighbors and wants to switch apartments with one of them. (In other words, each is jealous of the neighbors immediately above, below, or to the side, but not his diagonal neighbors.)

The question is: What is the fewest number of total moves that can accomplish this goal?

Teaching notes and solution: Again, give students time to talk to each other and work out a solution. Students probably have a sense by now that the puzzle is impossible, but they should be encouraged to express the solution in terms of parity.

One way to express the solution is this: if we label the squares A and B as in the diagram below, we can see that every A neighbor must switch with a B neighbor and vice-versa. But because there are thirteen A's and twelve B's, that can never happen. Notice that we don't need parity to state this problem, but once we express it that way (that the numbers of A's and B's have different parity so there can never be a one-for-one switch), we have a general solution that works for *any* n-by-n board, where n is an odd number.

A	B	A	B	A
B	A	B	A	B
A	B	A	B	A
B	A	B	A	B
A	B	A	B	A

Extension: Have students consider other arrays. They will see that for an m-by-n array, a solution is possible if either m or n is even, and the minimal number of moves needed (exchanges of apartments) is $mn/2$. Note that this number must be an integer because either m or n is even.

The Skywriter

Part 1: Show that the following squiggle could not have been made by a skywriter traveling from left to right in a single movement (without tracing over his own writing).

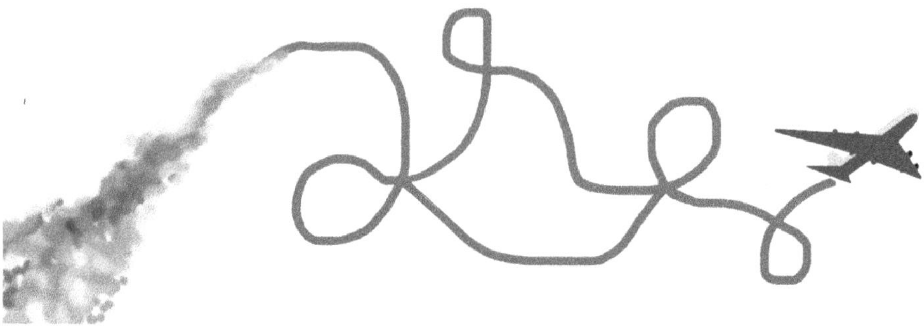

Teaching notes and solution:

Students will see that this is impossible, but, again, the instructor should challenge them to prove it. (It is not enough to say, "I can't do it – therefore it is impossible!") The discussion will probably focus quickly on the points of self-intersection of the skywriter's path, and students will see that the two intersection points where five lines meet cannot occur.

It may be difficult to articulate that the issue is one of parity. If the airplane goes into the intersection, it must come out as well. Hence the number of lines "going into" or "coming out of" the intersection point must be even. Since some of them here are odd, the path is not possible.

Couched as a skywriter's path, this problem is not hard to solve. The more general, and more useful, problem is about *graphs*, not skywriting paths. Without going into much detail, we can tell students that a graph is simply a set of lines and points. The lines are called *edges* and the points are called *vertices*. The problem is to create the graph as a skywriter would: by tracing it in a single path, without lifting the pencil off the paper, and without retracing any edge you've already traced.

Here are some examples. Unlike the situation with the skywriter problem, here students are not given a starting or ending point. They must find each for themselves.

Students find these problems intriguing. They often try to memorize a way to draw the tricky ones, thinking that the point is simply to solve the specific problem in front of them. If this happens, try to move them on to another problem. Sometimes just adding a single edge (line segment or curve) to the graph will be enough.

Students can also create their own problems. Have students draw their own graphs (with clear intersection points so that their work can be checked) and classify them as "possible" or "impossible." Ask students to think as they work about what makes one set of graphs different from the others. Give them the hint that it is a parity problem.

Notice that with graphs, as with the skywriter problem, every time a continuous path intersects itself, it creates a "vertex" with an even number of "edges" coming out of it. The number must be even because if the path continues, each line that enters a vertex must also exit it. In general, a graph with more than two odd vertices cannot be traced over with a continuous stroke without retracing part of the path. And if there are two odd vertices, one must be the starting point and one must be the ending point.

This set of problems is classic and turns out to be very useful. For much more on the topic, including interesting historical background, look up "The Seven Bridges of Konigsberg" or "Euler Paths" on the Internet.

The Sum-of-Fifteen Problem

Place plus signs and/or minus signs in the blanks to make a sum of 15.

8 _____ 7 _____ 6 _____ 5 _____ 4 _____ 3 _____ 2 _____ 1

Solution: The key is to realize that no matter how we place plus signs and minus signs, the parity of the total sum will not change.

This is easy to see when you work with a single pair of numbers at a time. For example, 8 + 7 is odd and 8 – 7 is also odd. Keeping this in mind, we can reduce the problem to the following slightly simpler problem where "8 ____ 7" is replaced by the word "odd" (since it must be an odd number):

odd _____ 6 _____ 5 _____ 4 _____ 3 _____ 2 _____ 1

We can work our way across, simplifying the problem a step at a time, and see that the total will be even no matter what combination of pluses and minuses we use. If so, then the sum can never be 15.

Extension: Have students make their own puzzle of this type. They will see that it is not quite as easy as simply changing the numbers. The sum of the numbers from 1 to 10, for example, is odd, but the sum of the numbers from 1 to 11 is even, and the sum of the numbers from 1 to 12 is also even.

Analysis of the Black and Red Card Problem

Have students play again through the Black and Red Card Problem from Day 2, and this time keep track of how many blacks and how many reds there are after each round.

For example:

	B	R	Total cards
Starting position	3	4	7
Turn over 2 R	3	3	6
Turn over 1 B 1 R	3	2	5
Turn over 2 B	1	3	4
Turn over 1 B 1 R	1	2	3
Turn over 2 R	1	1	2
Turn over 1 B 1 R	1	0	1

What is the invariant here? The number of cards reduces by one each time, but because of the way we replace the cards, the parity of the number of black cards is always odd. Since we end with one card of one color and zero cards of the other color, the last card must be black. (Remember that zero is an even number, so there cannot be zero black cards.)

Extension

Have students try to generalize the solution by experimenting with other configurations.

What about starting with three black cards and six red? We get the same solution for the same reason! It doesn't matter how many cards we start with. We could start with one hundred red cards, and the result would always be the same.

What happens if we start with three black and three red (both odd)? It doesn't matter whether there is an odd number of red cards. Notice that the parity of the red cards varies but the parity of the black cards is always odd, so black still must be the last card.

What happens if we start with four black and four red (both even)? This configuration is different and interesting. Black *cannot* be the last card this time because its parity is even and never changes. Red must be the last card because its parity changes, and it is the only one that can be odd. (And of course, some single card must be left, because the number of cards keeps decreasing by one.)

Example:

	B	R	Total cards
Starting position	4	4	8
Turn over 2R	4	3	7
Turn over 1 B 1 R	4	2	6
Turn over 2 B	2	3	5
Turn over 1 B 1 R	2	2	4
Turn over 2 R	2	1	3
Turn over 2 B	0	2	2
Turn over 2 R	0	1	1

Ginger's Pigeons: The Pigeonhole Principle

The pigeonhole principle is a principle of logic that can be applied to a surprising range of problems, often with surprising results. The problems can get very hard indeed, but we start off with very easy ones so that students can develop an intuition for the principle.

The pigeonhole principle seems at first simple and intuitive, so it is often a surprise to teachers that students find these problems difficult. The situation is compounded because the solution can seem obvious if it is read before the problem is worked. (This phenomenon is common to many mathematical problems but particularly vivid in problems using the pigeonhole principle.)

Students should be encouraged to work things out the long way as much as possible by using objects to represent the pigeons and holes (such as beans in cups), or by using pencil and paper to draw pigeons and holes, to see what kinds of arrangements are possible.

Introductory problems to work as a class

1. Ginger has four pet pigeons, and her pigeon house has four nesting boxes (called "holes"). Assume that every pigeon nests in a box and that the holes are big enough to fit multiple pigeons. If each pigeon nests in a hole, what can you conclude?

(a) Every pigeonhole must contain a pigeon.
(b) One pigeonhole contains at least two pigeons.
(c) Neither of the above.

2. Ginger's friend gives her a pigeon for her birthday. Now what can you conclude?

(a) Every pigeonhole must contain a pigeon.
(b) One hole contains at least two pigeons.
(c) One pigeonhole contains exactly two pigeons.

Solution to Problem 1: What possible configurations are there? We can't conclude very much. Any box could have from zero to four pigeons. The answer is (c).

Solution to Problem 2: What possible configurations are there? We could put all of the pigeons in a single box. So we certainly cannot conclude (a) or (c). The most we can conclude is (b), which seems like a pretty weak conclusion, but it turns out to be very useful.

The **pigeonhole principle**, simply put, states that *if more than* n *pigeons are placed into* n *pigeonholes, some pigeonhole must contain more than one pigeon.*

3. Mr. Kim has twelve students in his class, and each one has his own box of eight crayons. (There are the same eight colors in each box.) Each student reaches into his box, chooses a color, and writes his name at the top of his sheet.

What can we conclude?

(a) Somebody uses the red crayon.
(b) At least two students use the same color.
(c) At least two students use blue.

Solution to Problem 3: The "pigeons" here are the students, and the "holes" are the colors. Because $8 < 12$, we know at least two pigeons must be in a hole. So the answer is (b).

4. Ms. Wilkins has thirty students in her Algebra 1 class. They do not use crayons. What can we conclude?

(a) That at least two students must be hungry for lunch.
(b) That at least two students must be wearing the same color.
(c) That at least two students must have first names that start with the same letter.
(d) That at least two students must have been born in the same month.

Solution to Problem 4: We can conclude (c) and (d). In either case the students are the pigeons, and there are thirty of them. What are the holes, and how many are there? In the case of (d), the holes are the twelve months. Since $12 < 30$, we can conclude that at least two students must have been born in the same month. In the case of (d), the holes are the twenty-six letters of the alphabet. Since $26 < 30$, at least two students must share a letter.

5. *Did I ever tell you that Mrs. McCave*
Had twenty-three sons and she named them all Dave?
Well, she did. And that wasn't a smart thing to do.
You see, when she wants one and calls out, "Yoo-Hoo!
Come into the house, Dave!" she doesn't get one.
All twenty-three Daves of hers come on the run!
(from Dr. Seuss' *Too Many Daves*)

If the Daves share four bedrooms, then what can we conclude?

(a) Every bedroom has at least one Dave.
(b) One bedroom has at least two Daves.
(c) One bedroom has at least six Daves.
(d) Two Daves were born on the same day of the week.

Solution to Problem 5: We can certainly conclude (b), that one bedroom has at least two Daves. And we can also conclude (d) because there are more Daves than days of the week.

If students are ready for a more advanced discussion, in fact we can say more than that. If we spread the Daves out as much as possible in the four bedrooms we get:

$5 + 6 + 6 + 6$

So while we can easily conclude that one bedroom has at least two Daves, we can go further and conclude that one bedroom must have six Daves.

And in general:

If $n + 1$ objects are placed in n boxes, then one of the boxes must contain at least two objects.

If $nk + 1$ objects are placed in n boxes, then one of the boxes must contain at least $k + 1$ objects. This is a more general version of the pigeonhole principle, which we call the "strong version" of the principle.

Problems for Students to Work Alone or in Groups

From here, have students work in groups on the problems below and then come together to present their solutions to the class.

Level 1 Problems

In each case, solve (a) by identifying the pigeons and the holes, and (b) by showing that there are more pigeons than holes.

1. Eliza throws a die seven times. Show that she must have rolled the same number twice.
2. Scott draws six cards out of a standard deck of cards. Show that at least two of the cards must be of the same suit.
3. Show that if you circle thirty words on this page, at least two of them must start with the same letter.
4. Ian has a bowling party for his birthday and invites fifteen friends. Show that at least two of the friends must have knocked over the same number of pins on their first turn.

5. Kira's sock drawer has twelve blue socks, thirteen red, and seven green. She reaches in without looking. How many socks must she grab to make sure that she has a matching pair? How many must she grab to be sure that she has three matching pairs?

6. Carol has a pair of five-pocket jeans. If she puts a handful of thirty-four coins in her pockets, then what can we conclude?

   (a) Some pocket must have at least two coins.
   (b) Some pocket must have at least four coins.
   (c) Some pocket must have at least six coins.
   (d) Some pocket must have at least eight coins.

Level 2 Problems

1. Susanna randomly chooses books from her shelf to read. How many books must she read to ensure that she has read five books whose titles begin with the same letter?

2. Every one of the fifteen children in Peter's class has a lucky number. Each one likes prime numbers, and no one has a lucky number that is greater than his own age. Show that at least three children must have the same lucky number.

3. Show that at least two people in New York City must have the same number of hairs on their heads. (What information do you need to prove this?)

4. There are twenty people on the Smallville community softball team, and each one has a jersey with a two-digit number. Show that there must be two people on the team for whom the sum of the digits on the front of their jerseys is the same.

5. In Smallville Elementary school, there are 700 students. Show that two of them must have the same initials (for their first and last names).

6. Alek has a bowling party for his birthday and invites fifteen friends. If everyone in the room has at least one friend at the party, then show that two people must have the same number of friends in attendance.

Level 3 Problems

1. Show that in a major city (at least one million people) there must be at least twenty people who were born on the same date in the same year. (Assume people live to be no more than 120 years old.)

2. The lattice below consists of five columns of five points each, all equally spaced. Lines are drawn connecting pairs of lattice points, but never two in the same row or column. (An example is shown.) How many lines must be drawn to ensure that two of the lines are parallel?

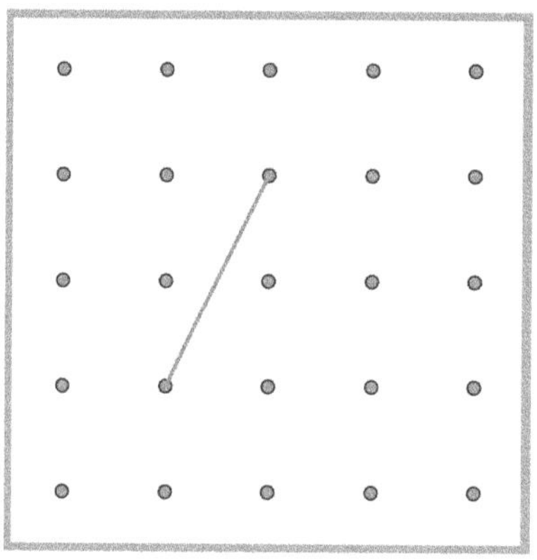

3. A huge deck of cards numbered 1 through 2014 is shuffled. Michael removes cards from the deck, one by one. What is the largest number of cards he must remove to ensure that two of the cards have a difference that is a multiple of 5?

Solutions to Pigeonhole Problems

Level 1 Problems

1. The six possible outcomes are the holes, and the results of the seven die throws are the pigeons. Since there are more pigeons than holes, Eliza must have thrown the same number twice.

2. The four suit possibilities (hearts, diamonds, spades, and clubs) are the holes, and the suits of the six cards Scott draws are the pigeons. Since there are more pigeons than holes, Scott must have drawn the same suit twice.

3. The twenty-six letters are the holes, and the initial letters of the thirty circled words are the pigeons. Again, since there are more pigeons than holes, two of the words must start with the same letter.

4. There are eleven ways to knock down pins (zero to ten pins) in one throw. These are the pigeonholes. The fifteen results of the friends' individual throws are the pigeons. Since there are more pigeons than holes, two of Ian's friends must have had the same result.

5. The holes here are the colors of the socks, and there are three of them. The pigeons are the colors of the socks Kira draws. To guarantee two socks of the same color, Kira must draw one more than three socks, so four socks. To guarantee three matching pairs, she must draw eight. Kira can have as many as seven socks and still only two pairs: B, B, B, B, B, R, G. With the eighth sock she must have a third pair.

6. The pockets here are the holes and the coins are the pigeons. Since $34 = (5 \times 6) + 4$, there must be more than six in some pocket. (In fact there must be seven in one pocket.) We can't conclude that there would be eight, however. To ensure eight coins in some pocket, we would need to have $(5 \times 7) + 1$ coins. Therefore (a), (b), and (c) are all true.

Level 2 Problems

1. $(26 \times 4) + 1 = 105$

2. There are 7 prime numbers less than 18: 2, 3, 5, 7, 11, 13, and 17. (We can restrict ourselves to numbers less than 18 because a child is younger than eighteen years. This is a conservative assumption. If we assume that a child is anyone younger than, say, 13 years, then our conclusion will still hold.) There are fifteen students in the class, and $15 = (7 \times 2) + 1$, so at least three must have the same lucky number.

3. The pigeons are the people in the city, and the holes are the different hair-count possibilities. We would need to know how many hairs a person *could* have on his head. It turns out that people have on average about 100,000 hairs on their heads, but we can assume twice as many just to be sure that our conclusion is sound. (That is, we increase the number of pigeonholes providing more space to distribute pigeons. If, in this harder scenario, we still have a pigeonhole with more than two pigeons, then we certainly will have a pigeonhole with more than two pigeons in the original case.) We also need to

know how many people live in New York because those are our pigeons. The United States Census Bureau estimated that there were 8,336,697 people living in the city in 2012. We can round it down to eight million, again rounding conservatively to make sure our conclusion is sound. (Here we are reducing the number of pigeons we distribute, again making the problem harder). Either way there are many, many more pigeons than holes. We could even conclude that there are at least forty people with the same number of hairs on their heads because eight million / 200,000 = 40. (We could fill our pigeonholes forty times over!)

4. The pigeons here are the players, and the holes are the possible jersey sums. The sum of the digits of a two-digit number could be anything from 0 (if we allow 00) to 18 (which is 9 + 9), so there are nineteen possibilities. Since there are twenty pigeons, at least two must have the same sum.

5. The pigeons are the people, and the holes here are the different possibilities for initials. There are 26 × 26 possibilities, which is 676. Since there are 700 students, at least two must have the same initials.

6. There are sixteen people at Alek's party: Alek plus his fifteen friends. (These are the pigeons.) Everyone must have from one to fifteen friends. (These are the holes.) There are more pigeons than holes, so two people must have the same number of friends.

Level 3 Problems

1. The holes here are the possible birthdates. There are fewer than 366 × 120 = 43,920 possible birth dates. (Not every year has a February 29, of course, but we don't need to be precise. We just need an upper bound for our figure.) There are one million pigeons, which is the population of our hypothetical city. 1 million / 43,920 is approximately equal to 22.8, so we can conservatively conclude that twenty people must have been born on the same date. Of course some birth dates will be much more common than others. (There are more forty-year-olds alive than hundred-year-olds.) But keep in mind that given our conservative assumptions, we have *proven* that twenty people have the same birth date.

2. Two lines are parallel if they have the same slope (the same ratio of rise to run). The pigeons are the lines, and the holes are the possible slopes. There are twenty-two possible slopes. (Since we are not connecting points from the same row or column, we don't have horizontal or vertical lines, and we don't have to deal with 0's anywhere.)

   1:1, 1:2, 1:3, 1:4, 2:1, 2:3, 3:1, 3:2, 3:4, 4:1, 4:3,

   -1:1, -1:2, -1:3, -1:4, -2:1, -2:3, -3:1, -3:2, -3:4, -4:1, -4:3.

   So if we draw in twenty-three lines, two of them must have the same slope. These two lines are parallel.

3. How many cards must we draw? Let's begin by considering the possible remainders when the numbers on the cards are divided by 5. There are five possible remainders: 0, 1, 2, 3, and 4. If two numbers have the same remainder when divided by 5, then they must differ by a multiple of 5. Since there are five possible remainders, we can choose as many as five cards, and it is possible that no two of them will differ by a multiple of 5. But if we draw a sixth card, its remainder must match with one of those we've already chosen. The answer is six cards.

The Mouse and Cheese Problem

This problem is good for both logic and spatial thinking. It also reinforces the idea of isomorphic problems (problems that are mathematically identical). Here students will be surprised that this difficult geometric problem can be solved by "borrowing" the solution from a simple and obvious problem.

Classroom Procedure

Have students work in groups of two to four. Each will need pencil and paper.

Tell students the following story:

"A mouse is exploring a house after a party, looking for leftovers. He climbs up on the table and finds a pile of twenty-seven little cubes of cheese. The cubes are stacked neatly and arranged into the shape of a large three-by-three cube. (Pause to ask the students what that looks like.)

The mouse eats the cheese by starting with one of the little cubes on the outside of the stack. Then he tunnels his way through each of the little cubes of cheese, one at a time. Each time he moves through one cube, he moves on to an adjacent cube. (Two cubes are adjacent if their faces lie flat against each other.) How can the mouse wind his way around, tunneling through each cube once, and ending with the center cube?"

Give students time to think about the problem, to ask questions and make sure they understand the problem, and to experiment with solutions. Students will model the problem in different ways on their paper. Some students may break the cube into two-dimensional layers to think about the problem.

If a group claims to have a solution, ask them to be sure that everyone in the group agrees and that everyone in the group can explain the solution. This gives the students an incentive to work together and also to clarify their thinking.

In fact, students will probably not reach a solution without an extra hint. Let the class work as long as the majority of the class is engaged. This might be fifteen to thirty minutes. Even if some students do reach a solution, do not have them explain it to the class. We want the class to reach its own solutions after the hint below.

The Hint: An Isomorphic Problem

Call the class together, and tell them that you are going to show them another puzzle, which will help them with the cheese puzzle.

Lay out twenty-seven cards on the floor, alternating black and red. There will be fourteen black cards and thirteen red.

Challenge the class to beat you in a game. Have students take turns coming to the front of the room to play where everyone can see.

The game: Tell the class that *yours are the black cards and theirs are the red.* (They will notice that you have more cards than they have. That's true, but they will get to choose who goes first.) The instructor and students will take turns picking up their own cards, one at a time. The goal is to be the last one to pick up a card.

The game can end two ways: If the student picks up one of his own red cards first, the game will end with two black cards. If the instructor goes first, picking up one of his black cards, the game will end with one black card. Either way, there is no way to end with a red card. You may need to play through a couple of times, but students will usually see this quite easily and many will be able to explain it. Give different students turns to explain the solution to this game in their own way.

Back to the Mouse and Cheese

Give the students the hint (if necessary) that they should color the twenty-seven cubes like the twenty-seven cards. Give students more time to work out a solution.

The key to the solution is realizing that if we color the cubes like a checkerboard such that adjacent cubes are always different colors, then no matter where the mouse starts, he is moving from black to red to black . . . (or red to black to red . . .). Regardless of how we choose to color the cubes (as long as they are in a three-dimensional checkerboard pattern) there are always thirteen of one color and fourteen of another. And the center cube is always part of the group of thirteen (the grey group in the illustration below), so just as we cannot draw a red card last, the mouse can never reach the center cube last.

Day 4

Goals for the Day

Students will be introduced to a powerful method of proof: mathematical induction. They will also revisit the concept of "isomorphism" by analyzing two games (Two-Row Nim and One-Piece Chess) that are mathematically identical. Students will also further explore the idea of "invariants" in the Leapfrog Activity.

One-Row Nim

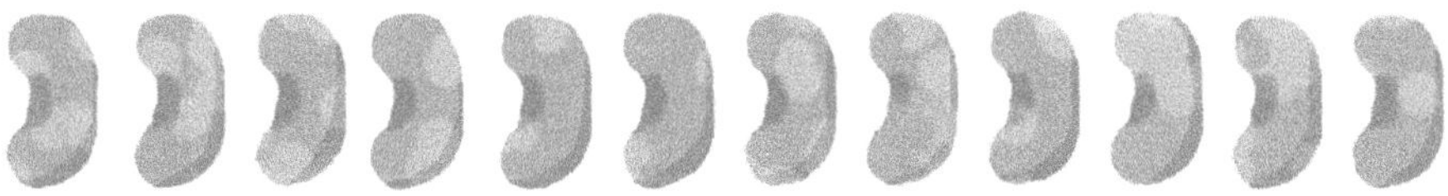

Nim is a name given to a whole set of very old games in which two players take turns picking up counters. There are many versions of Nim. Some versions limit the number of counters picked up in one move. Others divide the counters into separate piles with restrictions on how many piles a player can choose from.

Here we start with just one "pile" (it will be a row) of twelve counters (here they will be beans). Players take turns picking up beans. In one turn either player can take one, two, three, four, or five beans. The last player to take a bean loses.

The game is simple. It requires no background knowledge, yet it serves as a medium for discussing important mathematical ideas.

While there are lots of ideas embedded in this game, here we want to emphasize the informal use of mathematical induction.

Phase 1: Exhibition Game

Set up a demonstration game on the board and ask for a volunteer to play against the instructor. The instructor should ask the student who should go first and should let the student win. The class will learn the rules by watching the demonstration game. If the group needs more time to engage (this is typical with younger students), then play more exhibition games, having pairs of students play each other at the front of the room. Try to get as many students up to the front as possible. Many groups enjoy calling out suggestions for strategies as the game is played. This allows for greater participation, both numerically and emotionally.

Draw twelve beans on the board, and number them 1 to 12. (Students will play the game with real beans, which will not be numbered, but the game is easier to learn this way.)

1	2	3	4	5	6	7	8	9	10	11	12
O	O	O	O	O	O	O	O	O	O	O	O

Players take turns "picking up beans" in order from the right. In one turn a player must take at least one bean and can take as many as five beans. On the whiteboard, you take a bean by making a mark under it. It is convenient to record the players' turns this way so that students can examine them later and understand how the game was played. (In the illustration below, player 1 takes four beans and player 2 takes two.)

1	2	3	4	5	6	7	8	9	10	11	12
O	O	O	O	O	O	O	O	O	O	O	O
						/	/	×	×	×	×

The loser is the person who takes the last bean (bean #1). Ask students to imagine that the beans are jellybeans, and bean #1 is the bad jellybean, with an unpleasant flavor such as fish or hot sauce. The loser is the one forced to eat the bad jellybean.

Note: It is best to play the game with dried lima beans, or other large, flat beans. Using actual jellybeans often distracts the students from the game.

Phase 2: Students Play in Pairs and Look for a Strategy

The goal of this phase is to give the students time to think about the game on their own, test out ideas, and develop a strategy.

First, break students into pairs, and give each pair twelve beans. Instruct students to lay the beans out in a line and to play the game on their own.

Once it is clear that students understand the rules, ask them to find a strategy for winning the game. For students who need it, lead them with the following questions:

(1) Would you rather go first or second?
(2) How many beans do you want to leave your opponent with at the end? Leaving him with one bean would be good, of course. But what about leaving him with two beans? With three?
(3) Don't think so much about how many beans you should pick up. Concentrate on how many beans you want to leave your opponent.

The strategy that will emerge is to choose to go first and to take five beans in the first turn. One way to think about the strategy is that player 1 wants to leave player 2 with seven beans. If player 2 is left with seven beans, then he cannot take enough beans to leave player 1 with the last bean, and yet any amount he does take will leave player 1 with a winning move. As long as player 1 leaves player 2 with seven beans and does not make an error, he must win. We say player 1 has a "winning strategy."

You can help a student to see the strategy by asking what they would do if they started with eleven beans rather than twelve. Typically, they will see right away that they must take four beans, and this turns their attention to the number of beans remaining, which is the key to finding a general strategy in the next stage.

If two students in a pair do not seem to be progressing in developing strategy, split them up by pairing each with a student who has more insight.

Phase 3: Students Play With More Beans and Look for a More General Strategy

Different pairs of students will move on to phase 3 at different times, and some may never reach this phase. As soon as a pair verbalizes the strategy for the twelve-bean game, they should begin this phase and look for a more general solution.

Students who stop before phase 3 are in danger of simply memorizing the rule for winning this particular game. The point of this activity is of course not that. It is to get the students to think about strategy in general. So let the students enjoy their mastery of the game, but move them on as soon as possible.

Give the pair three more beans, making the total fifteen in all. Ask students to figure out a strategy for this "new" game.

The general strategy:

Recall from the twelve-bean game that the loser is the one who is left with seven beans. No matter how many beans he takes, his opponent can force him to take bean #1.

The key to solving the fifteen-bean game is that the position with seven beans is still a losing position, even if more beans are added. Bean #7 remains "secretly poisoned." Students often learn this idea intuitively, but it is worth articulating in a class discussion after most of the class has solved the fifteen-bean game.

In fact, this pattern continues with every sixth bean being secretly poisoned. If a player can force his opponent to take bean #13, then he can force him to take bean #7. If he can force his opponent to take bean #7, he can force him to take bean #1. In general, the player who picks up a bean numbered 1, 7, 13, 19, etc. ($6K + 1$) is in a losing position. So, by taking every bean down to a bean numbered $6K + 1$, the first player can force the second player to take a "bad" bean and therefore can win. (See the example game below, where player 1 takes two beans in his first turn, forcing player 2 to take bean #13, then uses that position to force his opponent to take bean #7, then bean #1.)

1	2	3	4	5	6	7	8	9	10	11	12	13	14	15
O	O	O	O	O	O	O	O	O	O	O	O	O	O	O
/	X	X	X	/	/	/	X	X	X	X	/	/	X	X

At this point the instructor must judge when the group is ready to talk about the experience. Some faster students can work with twenty beans, just to see if they understand the game in full generality. Even if they do, giving them practice with more beans will enrich the discussion later.

Phase 4: Class Discussion

Call the group together to discuss their experiences and what they have learned.

Discuss the solution to the twelve-bean game, which most groups will have solved. If students say that the strategy is to "go first and take five," then use this as an opportunity to have them describe the strategy differently. We want students to get to the idea that leaving the opponent with seven beans puts him in a losing position. That is, we want to help them see that if a player takes bean #7, no matter what he took before that, and no matter what other bean he takes, he will lose.

If this point does not come up naturally in conversation, there are several ways to bring it out. One is to play a few games and circle bean #7 in red. Students will note that taking bean #7 makes a player lose. Or, if necessary, the instructor can call their attention to the record of exhibition games and point out that bean #7 was always taken by the loser. It is easier to see this pattern if the games are recorded in rows on the board.

One effective way to describe the danger of taking bean #7 is to continue the food metaphor. Bean #7, like bean #1, is "poisoned" with a bad flavor, but secretly so.

It is a good idea to ask the students to verbalize why this is true, even if they all see it. You can then elicit the idea that the player who forces the other to take bean #7 always has a winning "reply" to the move. This is another element of strategy, and making it conscious will help students use it in new contexts.

If enough of the students have had experience with fifteen beans, you can launch a discussion of the general strategy right away. The object of the discussion should be that bean #13 is also "secretly poisoned." If a player is forced to take bean #13, then his opponent can force him to take bean #7, and thus to lose.

Once students have this insight, the game becomes abstract. They no longer need beans and can think about the essential elements of the game without the physical props.

So, the instructor can now work with those students who are ready to advance. They can be asked to imagine a row of one hundred beans, and to identify the numbers of the "poisoned" beans. They typically come up with a list:

1, 7, 13, 19, 25, 31...

and a rule: "Add 6 to get the next poisoned bean." This is a *recursive* rule.

(With a recursive rule, to find the value of any given term, we must know the value of the previous term and therefore must compute all previous terms.)

A good question to ask at this point is whether the first player always wins, no matter how many beans are on the table initially. Many students will answer "yes," then realize that if there are 7, or 13, or 19 beans on the table initially, the first player must lose, not win. This, too, is an important step in the formulation of the concept of a winning strategy.

But what if they are presented with 1000 beans? Should they go first or second? How should they play? A story about such a game, say with a very powerful opponent (the devil), might be good motivation. The issue is that to avoid counting by sixes all the way to 1000, students must find a pattern to the poisoned numbers that does not depend on the previous poisoned number. They must eliminate the recursion.

One way to get students to see how to eliminate the recursion is to list the bad beans vertically as in the example below, then have a student subtract 1 from each of the poisoned numbers and place it to the left of the list of bad beans:

	Bad Beans
0	1
6	7
12	13
18	19
24	25
30	31
36	37

Students in grades 4 or higher will quickly recognize the "6-times table." At this point, the instructor can introduce some algebraic notation. (Don't be afraid to do this!) Many students will understand, and those who don't can pick up the thread of the argument a bit later, without the algebraic notation. At this point in the activity, most students will have in mind the concept of division with remainder, and its application here. The algebra simply documents what they have thought of.

The numbers in the times table are all "6 times something," and most students will understand that they can be represented as 6K. So the poisoned numbers are of the form 6K + 1.

For 1000 beans, then, we must find the multiple of 6 just below 1000, and then add 1 to find the first poisoned bean. Students who don't see the algebra can understand this formulation and are likely to see that the tool they need is division with remainder.

Now division with remainder means finding the values of Q and R in the equation 1000 = 6Q + R, where Q is the quotient and R the remainder. But most students will not think of it in this algebraic notation. They will probably perform the division as they were taught to:

$$\begin{array}{r} 166\text{ r.}4 \\ 6\overline{)\,1{,}000} \end{array}$$

Ask students to check the division. This will give them the usual algebraic form: $1000 = 6 \times 166 + 4$.

It is good for students to multiply out $6 \times 166 = 996$ to see that this works. Many will have been trained to do this anyway.

You can point to the lists of numbers, to emphasize that 996 is in the list of multiples of 6. Then bring them back to the game. Bean #997 is the first poisoned bean, so they want to be player 1 and to take three of the 1000 beans, leaving their opponent with bean #997.

Extensions

This game is very rich. If students show interest, ask how they can change the rules of the game to make it more interesting. We can vary not only the number of beans on the table, but also the number of beans a player is allowed to take on a turn.

In general, the game can be described as starting with N beans, and players choose from a set S of moves, which is a set of numbers of beans that can be removed in one turn. For the initial game, $N = 12$ and $S = \{1, 2, 3, 4, 5\}$.

The game has not been solved in full generality, but only for particular values of N and S. So it may not be a good idea to choose S arbitrarily. Some useful variants are as follows.

1) $S = \{1, 2, 3, 4\}$. Here the poisoned beans are $5K + 1$, and it will be easy for students to describe the numbers as those ending in a digit 1 or 6.

Usually, this is not a good variation for the first game. Students may focus on this description and will not be motivated to go on to describe the poisoned beans using division with remainder. But as a second game, it is good reinforcement and offers two valuable insights: (1) the last digit of a number (in decimal notation) is the remainder when the number is divided by 10, and (2) if we know the remainder when a number is divided by 10, then it is easy to find the remainder when the number is divided by 5.

2) S = {1, 2, 3, 4, 5}; N = 12 (the initial game), except that the winner is the person who takes the last bean, not the one who avoids it. A "poisoned bean" strategy again emerges. This time the poisoned beans are simply those of the form 6K.

Connection to Mathematical Induction

Mathematical induction is a method of proof that is used often in mathematics, and the game of Nim serves as a concrete example of its application. The idea of mathematical induction can be touched on or discussed deeply, depending on the group of students. Either way, begin by asking students, "How do we use logic in finding a strategy?" Students will have lots of ideas about this. Here we want to make explicit that a game with a large number of beans can be reduced to a game with a smaller number of beans.

That is: how did students figure out that every 6K + 1 bean is poisoned and that it continues to be poisoned even when the numbers get really big? One way to describe the thinking is as follows:

1) In the twelve-bean game, bean #7 is poisoned. So we might as well have covered up the first six beans, and played as if the goal were to make your opponent take bean #7.

2) We can use this insight to solve the game with fifteen beans. We can pretend we are playing for bean #7 (which is poisoned) instead of bean #1. So let's forget about the first six beans. We are now playing a game with nine beans (15 – 6 = 9), and we know that bean #13 is poisoned because in this truncated game, it is now the seventh bean. So the first player can win by taking two beans.

Rephrasing the thinking like this will eventually help students get to the idea of mathematical induction. And for now, the notion that we can reduce a problem to a smaller problem is a big step.

There are two more cognitive steps to take if students are ready, which are outlined below.

The first step is to think just about the poisoned beans. Because of the rules of the game, the fact that bean #7 is poisoned *implies* that bean #13 is poisoned. And the fact that bean #13 is poisoned likewise implies that bean #19 is poisoned, and so on.

Then the argument, by mathematical induction, is as follows:

a) We know that the first bean is poisoned.

b) We know that if bean M is poisoned, then bean M + 6 is also poisoned.

By statement (a), the first bean is poisoned.
Therefore by statement (b), bean #7 is poisoned.
Therefore by statement (b), bean #13 is poisoned.
Therefore by statement (b), bean #19 is poisoned, and so on.

So all the beans of the form M + 6K are poisoned, where M = 1 and K = any integer. This is a proof by mathematical induction.

A common metaphor for mathematical induction is the classic row of falling dominoes. How can we prove that all the dominos in a line will fall? We must show 2 things:

(1) That the first domino is knocked over. (In other words, that the statement is true for some starting value.)

(2) That IF a domino falls, THEN the next domino also falls.

You can use actual dominoes to represent this as the class discusses it. However, we have found that it is a useful metaphor only after students have some idea what is happening in another context, such as the Nim game above.

Two-Row Nim and One-Piece Chess

These two games are related mathematically and are best played one right after the other so that students can see their relationship.

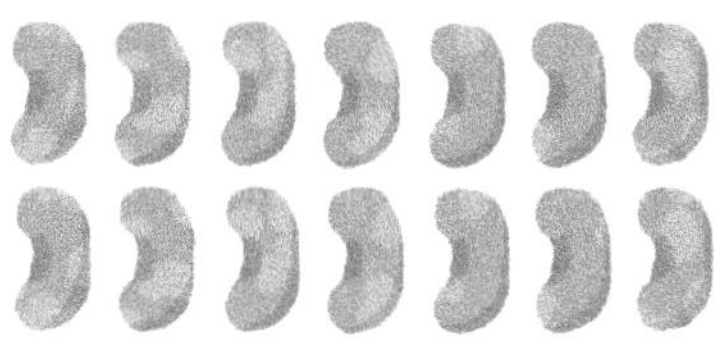

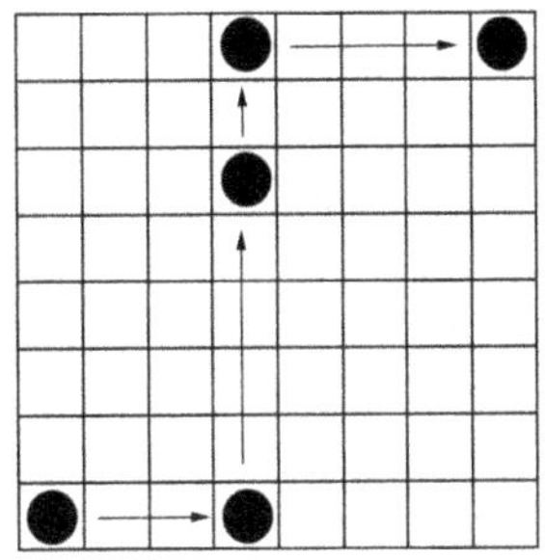

Like One-Row Nim, these are two-person, perfect-knowledge games. In all of these games both players have complete information, unlike many card games in which some information is hidden. The games involve only strategy, not luck or chance.

The notion of strategy does not always come easily to students. They often think step-by-step: "If my opponent does this and this, I will reply with such-and-such a move," but without a notion of generalizing the play.

Two-Row Nim

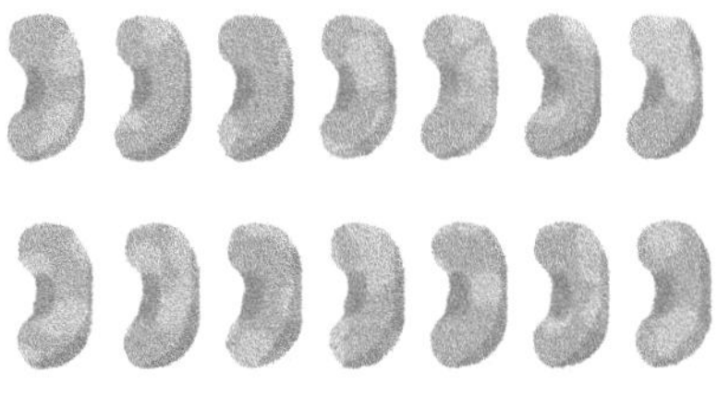

Phase 1: Exhibition Game

Set up a demonstration game on the board and ask for a volunteer to play against the instructor. The instructor should ask the student whether he wants to go first and should let the student win. Watching the demonstration game will help the class understand the rules.

Draw fourteen beans on the board in two rows of seven.

Players take turns "picking up" beans in order from the right. (Picking from the right is not essential to the game, but makes it easier to strategize if players remove beans from the same end each time.)

In one turn a player must take at least one bean and can take as many as he likes from either row. However, in each turn a player can only take beans from a single row.

The figure below shows an acceptable first move. Here, player 1 has taken five beans from the top row. Player 2 can respond by taking one or two beans from the top row or taking up to seven beans from the bottom row.

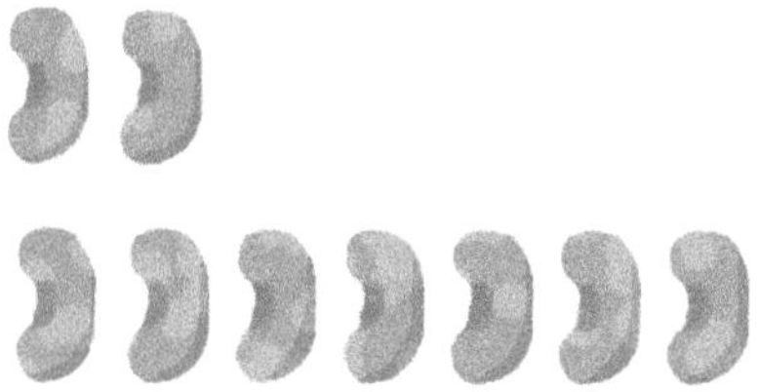

The winner is the player who takes the last bean.

Phase 2: Students Play in Pairs and Look for a Strategy

The goal of this phase is to give the students time to think about the game on their own, test out ideas, and develop a strategy.

First, break students into pairs, and give each pair fourteen beans. Instruct students to lay the beans out in two rows of seven and to play the game on their own.

Once it is clear that students understand the rules, ask them to find a strategy for winning the game. The strategy that will emerge is to choose to go second and to copy all of the other player's moves. That is, however many beans your opponent takes from one row, you take the same number from the other row.

One way students often develop a strategy is by working backwards. How does the game end? This is a good hint to give students who are stuck.

You can ask them: Is this a good position to be in?

Clearly not. I can take only a single bean, and I leave my opponent with the last one. We say that this is a "losing position." In that case, the position below must be a "winning position" because from this position I can take one bean and put my opponent into a losing position:

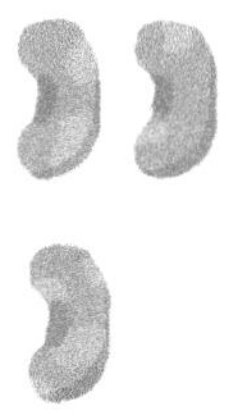

In general, a losing position will have rows with the same number of beans, and a winning position will have two rows with different numbers of beans. Because the game begins with two rows with the same number, the strategy is to let one's opponent go first and to copy all of his moves, evening out the rows with each move and thereby putting him in a losing position each time.

In this game it is not a good idea to pair faster students with slower students. Once a student sees his opponent use the winning strategy, he will recognize it immediately, and the exercise of deriving it for himself will not be possible.

Phase 3: Class Discussion

Call the group together to discuss what they have learned.

Discuss the solution to the game. Students may say that the strategy is to "go second and copy the other person," which is a good start, but it is important to get them to move towards a more general strategy.

Ask: what if the game started with one row of five and a second row of nine? What would the strategy be then? This time you wouldn't want to go first. The idea is to get students to focus on what makes a winning position and a losing position.

One way to get students to reflect on the strategy is to play another game on the board and ask the class to articulate the reasons for their moves as they go. You want them to be able to say clearly that evening out the rows hands your opponent a losing position because you can always force your opponent, move by move, into the ultimate losing position, with a single bean in each row.

One-Piece Chess

A chessboard (sixty-four squares) can be photocopied or drawn freehand. There is no need to color the squares black and white. It is easy to draw a chessboard by dividing a square into quarters, and then dividing each of those quarters in half vertically and horizontally (to make sixteen squares) and then dividing each of those in half again vertically and horizontally (to make sixty-four squares). This in itself is a worthwhile activity.

Phase 1: Exhibition Game

Set up a demonstration game on the board and ask for a volunteer to play against the instructor. The instructor should ask the student whether he wants to go first and should let the student win.

Watching the demonstration game will help the class understand the rules.

Set up an eight-by-eight grid on the board. Draw a rook (a circle will do) in the bottom left-hand corner.

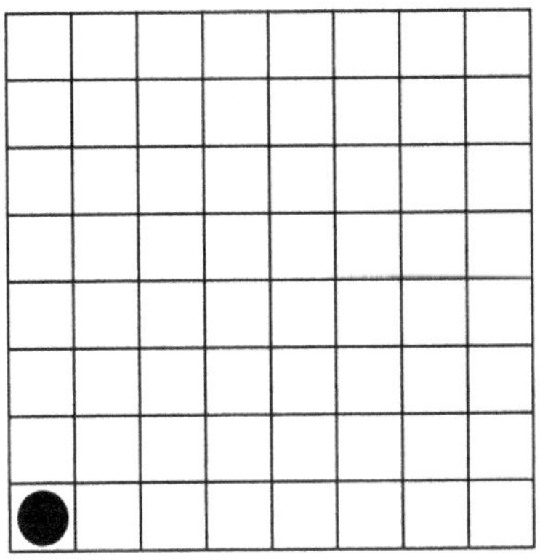

The rook can move only up or to the right, never down or to the left, and never diagonally. Both players move the same piece, taking turns. They can move it any number of squares up, or any number of squares to the right (but only one direction in a turn). The goal is to move the rook to the upper-right corner.

In the exhibition game, each player should show the movement of the rook on the board as in the example below. (When players play on their own, they will move a counter on the chessboard without keeping track of its history.)

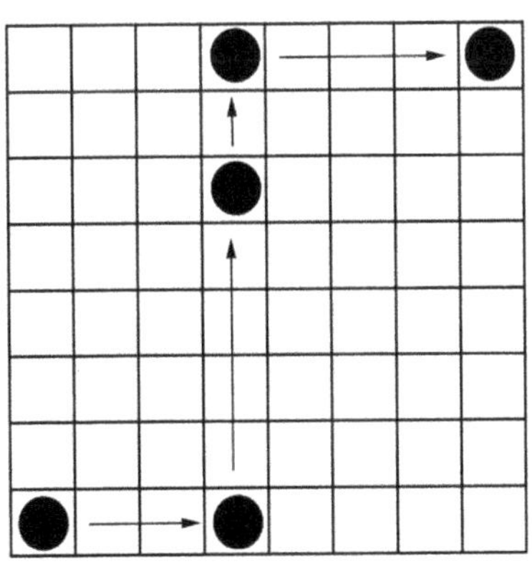

Phase 2: Students Play in Pairs and Look for a Strategy

The goal of this phase is to give the students time to think about the game on their own, test out ideas, and develop a strategy.

Break students into pairs to play on their own. Give each pair a piece of paper with an eight-by-eight grid and a bean or other counter.

Once it is clear that students understand the rules, ask the students to find a strategy for winning the game.

The strategy that emerges is to choose to go second and to respond to the first player's move always by moving the rook back onto the diagonal. Player 2 has a "winning strategy."

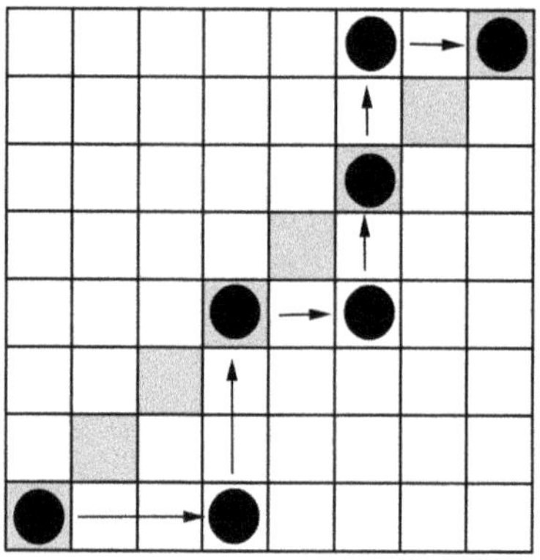

This strategy typically emerges from the play. Students see that finding the rook on the seventh diagonal square (counting from the bottom) is a losing position: the player whose turn it is must move off the diagonal, and then the other player can win.

The next insight, typically, is that the sixth diagonal square is also a losing position: from there any move you make will permit your opponent either to win in one move, or to move the counter onto the seventh diagonal square. But if the sixth diagonal square is a losing position, then so is the fifth, the fourth, and so on.

For "experts," help them to generalize by altering the game board. Once a pair has solved this game, the instructor can fold over three rows of the chessboard, and ask them what would happen if they started with a five-by-eight chessboard. In this situation, the first player has a winning strategy because he can move onto a "diagonal" (i.e., a diagonal of the five-by-five subset of squares that includes the last square). Essentially, the roles have been reversed.

Here, as in the case of Two-Row Nim, it is not a good idea for slower students to play against faster students. It is too easy for the slower student to note and memorize the strategy without having a chance to figure it out for himself.

Phase 3: Class Discussion and Connection to Two-Row Nim

Call the group together to discuss their experiences and what they have learned.

Discuss the solution to the game. Students will see that the strategy is to go second and move onto the diagonal.

For advanced students, a discussion might ensue about what a winning position is. Formally, a set S of winning positions is such that

(a) the last position in the game is in S;
(b) from any position outside of S you can move into S; and
(c) from a position inside S you must move outside S.

The set of diagonal squares here satisfies these conditions. The first player starts at a position in S, and must move to a position outside S. The second player then has a winning strategy by always moving to a position inside S.

But it is not necessary for students at this level to make this generalization. They will still have learned from the game.

This game is *isomorphic* to Two-Row Nim, and this fact is the real reason to play these two games together. "Isomorphism" is an important mathematical concept. It means, essentially, that the two games are mathematically identical. If you know how to win one of them, you know how to win the other. In the chess game, there are seven possible moves upwards and seven possible moves to the right. We can think of each move a student makes as taking away that many moves from the possible moves of an opponent upwards or to the right, the same way we take away beans in Two-Row Nim. In fact, when we play Two-Row Nim, we can keep track of the moves in the corresponding chess game with two rows of beans, the "upwards" beans, and the "to the right" beans.

Students get this idea quickly if you play an exhibition chess game with one of them, and record the moves simultaneously in "Nim" fashion.

For experts: Three-Row Nim is much more complicated. It can be envisioned as a three-dimensional array, but it is harder to understand what the winning positions are. The diagonal does not play quite the same role in three dimensions. The game has been solved, and the solution is accessible to middle school students. But it's the sort of thing they will memorize as a trick, rather than learn from.

Leapfrog Activity

This is a puzzle that can be worked with pairs or groups of students. It takes about forty-five minutes to an hour to work through the puzzle and then discuss as a class. The puzzle requires a grid or lattice, such as is often formed by square tiles on a floor. If you have this opportunity, students enjoy working as a group on the floor.

Phase 1: Exhibition

Clear a space on a tile floor where all students can see.

Put three counters on three vertices of one of the tiles. These counters are "frogs," and they move by playing leapfrog.

This is the starting position:

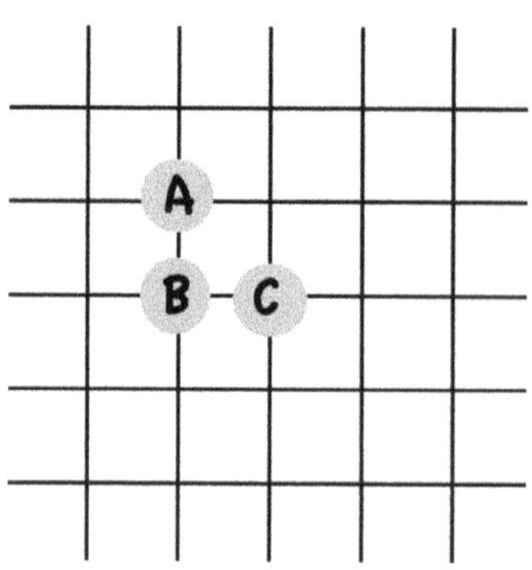

Start by explaining the rules for moving frogs, and let students experiment.

The rules: Frogs can move only by jumping over another frog. When a frog moves, it is reflected over the other frog, landing on a new vertex. That is, frog A can move to a point for which some other frog B sits at the midpoint between frog A's old and new positions. (The geometric transformation describing this is *point reflection*. It is fully equivalent to a rotation by 180 degrees about the position of frog B, but the equivalence is not obvious, and not useful for playing this game.)

Before telling students the goal, ask them to take turns moving frogs. With each example students will get a better understanding of which moves are allowed.

For example: Here frog A jumps over frog B to a new position:

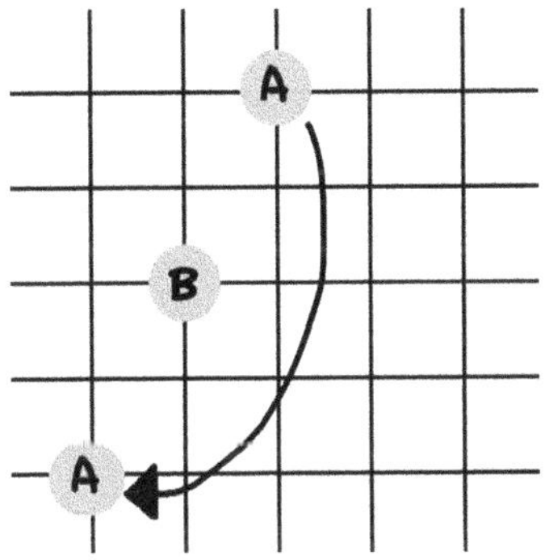

Place frogs back in their original positions and instruct students that the goal of the puzzle is to get any one of the three frogs to land on the fourth vertex of the original square. Mark that vertex in some way, such as by putting tape over it in the shape of an X.

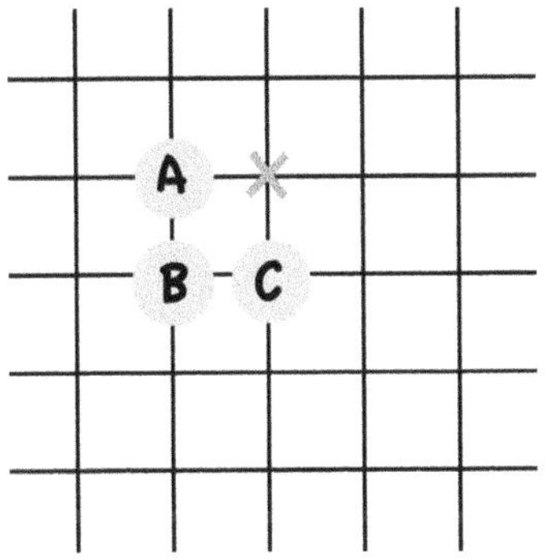

One way to motivate the game with younger students is to tell a tale:

An evil sorcerer turned three young wizards into frogs. Imprisoned on a lattice, they can only move as described above. The fourth vertex of the square is a portal that allows them to escape from their lattice prison.

Unfortunately, as we shall see, there's no happy ending. So use your discretion in telling this tale.

Phase 2: Dividing Students up and Setting up the Puzzle

Students can work in groups of two to four. Divide students up and give them grid paper to work on, or assign them a large section of tiled floor. Let students work as long as they are engaged. In fact, it is not possible for a frog to reach the X, and students may get this sense. If they start suspecting this, ask them to think about why. How could they prove it?

Phase 3: Class Discussion

Call students together and play the game once more together, this time on the board. Ask students what they noticed about the movement of the frogs.

We can show that it is impossible to reach the X if we introduce coordinates on the lattice so that the frogs start at (0,0), (0,1), and (1,0). Then the parity of each frog's coordinates never changes. (It is *invariant.*) And since none of the frogs starts at a point with two odd coordinates, none can ever land on (1,1).

Students can find this out by keeping track of those points where each frog lands. That is, suppose each frog has wet paint on its feet, each a different color. Students can keep track of where each frog can land by coloring the vertices where each lands as they experiment. They will find out that each frog moves on its own sub-lattice, the lattice of points with coordinates of the same parity. There is a fourth sub-lattice on which no frog ever lands, and the X is on that sub-lattice.

It takes some experimentation to see the sub-lattices, but once students see the visual pattern, they accept that it is impossible for a frog to land on the X. The teacher must decide based on the students' level of understanding whether to explain the solution in terms of the invariant parity of the coordinates.

Classroom Hints

Sometimes two groups of students can combine their work. That is, if they are keeping track of where each frog lands, they can pool their work by extending their lattices until they join together. But the teacher must take care that this is done correctly, with the corresponding sub-lattices fitting together.

For Experts

Notice that the triangles formed by the three frogs always have the same area. If we start with triangle ABC, and reflect point A over point C to get point A', then the areas of triangles ABC, A'BC are equal: they have equal bases (AC = A'C) and the height of point B above line AC does not change. Since the area of a triangle depends only on a base and the corresponding height, the area does not change. Thus the area of the triangle formed by the frogs is *invariant.*

The game can also be played starting with three frogs on three arbitrary (but non-collinear) points. The frogs' moves then generate a lattice of parallelograms rather than squares, and almost all the observations made above still hold.

Day 5

Goals for the Day

Students will gain more experience with logical deduction, including complex chains of reasoning in the Hidden Card Game and reasoning about numerical relationships in the Magic Squares Activity. In the Girls and Boys Problem, students will gain more experience with invariants. In the relating the Magic Squares Activity to the Fifteen Game, students will also see another example of an isomorphism.

The Girls and Boys Problem

The Problem

Four boys and four girls stand in a line. At a given signal, the boys all turn right (say, to the door) and the girls all turn left (say, to the window). Each boy peers ahead and tells how many students he sees in front of him. Then each girl does the same thing. These numbers are recorded, and the boys' and girls' "scores" are added up. The boys and girls take turns organizing themselves to maximize their score.

For example, here's how we would score an arrangement:

Girls: 0 + 3 + 4 + 6 = 13
Boys: 6 + 5 + 2 + 0 = 13

(A tie!)

Classroom Procedure

The instructor explains the problem and then brings eight students to the front of the room to act out the problem.

The instructor should choose one student to keep score at the board. The students in line can report their individual "scores" to the scorekeeper, and the scorekeeper can record the information on the board using the format below.

	Girls	Boys
Round 1	0 + 3 + 4 + 6 = 13	6 + 5 + 2 + 0 = 13
Round 2		
Round 3		

The students who are seated and not participating actively should be encouraged to give advice to the students at the front of the room, to double-check the scores reported by individuals, and to double-check the scorekeeper's work.

After the first round (in which the students lined themselves up randomly), the girls' and boys' teams should take turns choosing an arrangement for the line, each trying to find an arrangement that will outscore the other team.

As they rearrange themselves and add up their scores a few times, students will notice that *for any arrangement, the boys' and girls' sums will be equal.*

At this point students can think about the problem abstractly, without acting it out. Students should take a seat and work in groups to discuss the following questions. (Give the questions one at a time.)

1. You have noticed that there is always a tie, but is the tie score always the same? Explain.

2. In fact the boys' and girls' scores vary, but the difference in the teams' scores remains constant. Why must that be true?

3. How can you maximize the boys' (or girls') score? How can you minimize it?

Analysis

1. You have noticed that there is always a tie, but is the tie score always the same? Explain.

The tie score is not always the same. It varies, and students can prove this by example.

2. In fact the boys' and girls' scores vary, but the difference in the teams' scores remains constant. Why must that be true?

The difference of the boys' and girls' scores will always be zero. The difference is *invariant.*

There are different ways to analyze this problem and prove that the difference is always zero, and students will have their own explanations. The teacher will have to follow their words closely. Some student explanations amount to simply describing the phenomenon and not really showing that it must be true. While students often do not have the background to offer a formal proof, with guidance they will be able to come up with, or at least follow, the reasoning described below.

One way to prove that the difference in scores must be zero is to show
(1) that some particular arrangement gives a difference of zero and
(2) that any switch of two students will add (or subtract) an equal amount to both the boys' and girls' scores.

Suppose we start off with the boys on one side and the girls on the other in a symmetrical arrangement. Certainly (by symmetry) the difference should be zero. But let us check:

Girls: 0 + 1 + 2 + 3 = 6
Boys: 0 + 1 + 2 + 3 = 6

(Another tie!)

Now suppose we switch two children randomly. If we switch two students of the same gender, then nothing at all changes:

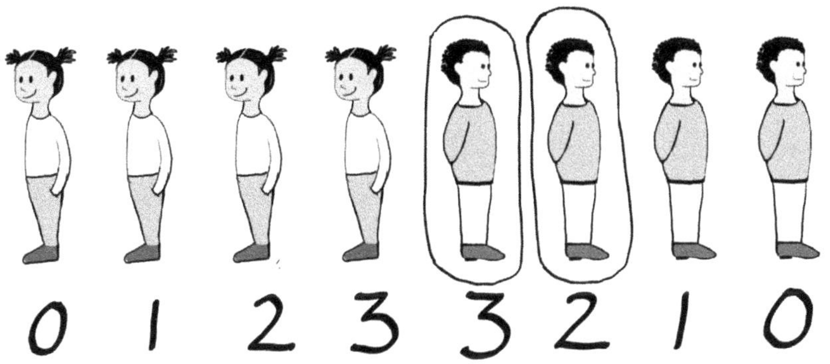

Girls: 0 + 1 + 2 + 3 = 6
Boys: 0 + 1 + 2 + 3 = 6

(Another tie!)

What happens if we switch a boy and a girl?

Girls: 0 + 2 + 3 + 4 = 9
Boys: 6 + 2 + 1 + 0 = 9

(Surprise! Another tie!)

Notice that when we switched the boy and the girl, it had the effect of moving a boy back three spaces and, at the same time, moving the girl back three spaces, thereby adding three to each team's score. In fact, this is true in general, and students can convince themselves by acting it out a few times. Any boy/girl switch has the effect of moving a boy n spaces in the line and moving a girl move n spaces in the opposite direction, thus altering the scores but keeping the *difference* equal. The difference remains invariant.

Notice that we haven't shown that *any* arrangement of boys and girls can be created by starting with a symmetrical arrangement and making a series of individual boy/girl switches. Our "proof" depends on this, and for now we just assume it to be true. But that is another interesting topic of discussion.

3. How can you maximize the boys' (or girls') score? How can you minimize it?

The maximum score is 22:

The minimum score is 6:

Notice that both solutions are symmetrical. (Why this happens is not easy to explain.)

Other points to consider:

4. What happens if there are more girls than boys? Certainly the score won't always be tied. But can the score ever be tied under this condition? Is the difference still an invariant?

5. Can you arrange the boys and girls to make a certain target score, such as 10? Is every score possible to make between the minimum and the maximum scores?

The Hidden Card Game

Students sit in a circle facing each other. (There should be at least five.) They are each given a card, which they must not look at. Each student places the card on his forehead so that others can see it but he cannot. In each round the instructor gives the students some piece of information about the number of black and red cards. (Be sure to write this on the board.) The instructor then asks students whether they know the color of their card based on the colors of the cards that are visible to them. Students who do should silently raise their hands without saying anything. In particular, a student who knows the color of his card should not tell what he thinks the color is. The fact that he *knows* his color is the important piece of information he shares with the group by raising his hand.

After each round it is important to let students talk about what they were thinking and how they figured things out. It is also a good exercise for them to explain others' thinking: "How did Carol know hers was black?"

If the logic is difficult for the students, the class can play each round multiple times. Once one round becomes easy for them, then move on to the next one.

Round 1

What to do: Give out one red card and the rest black.

Instruction to students: "There is exactly one red card. Raise your hand if you know your color."

Analysis: This is enough information for everyone to know his color right away. Those who see one red card know they have black. The one who sees no red card knows he has red.

Round 2

What to do: Give out three red cards and the rest black.

Instruction to students: "There are exactly three red cards. Raise your hand if you know your color."

Analysis: Again, this is enough information for everyone to know his color right away. Those who see two red cards know they have red and those who see three red cards know they have black.

Round 3

What to do: Give out one red card and the rest black.

Instruction to students: "There is at least one red card. Raise your hand if you know your color."

Analysis: The student with the red card should be the only one to raise his hand at first. He can see that there are no other red cards, so his must be red. Other students may *then* realize from his reaction that theirs must be black. (Specifically, they see that he was the only one to raise his hand at first and therefore that he must see all black cards.)

Clearly, wait time is important in this round, as is silence on the part of the students.

Round 4

What to do: Give out two red cards and the rest black.

Instruction to students: "There are either one or two red cards. Raise your hand silently if you know your color."

Analysis: With the more complex versions of the game, students deduce things based not only on what they see, but also on how other students answer. The students with the black cards should be the only ones to raise their hands at first. They can see that there are two red cards, so they know there are no others and so theirs must be black. The students with the red cards may then react to the others and conclude that theirs must be red.

This version can be played with different numbers, and for some groups of students, this may be sufficient challenge for the entire activity. For example, give out three red cards and tell them that there are either one, two, or three red cards.

Round 5 (advanced)

What to do: Give out two red cards and the rest black.

Instruction to students: "There is at least one red card. There could be one, two, three, or more. Raise your hand silently if you know your color."

Analysis: At first no one will raise his hand. From that students can conclude that there must be more than one red card. (If there were only one red, then that person would have raised his hand.) Once students see that there must be more than one red card, the ones with red cards will know that theirs are red (because they will see only one other red). From there, everyone will know his color.

Extension: You can play the same game but with three red cards or even four. The situation rapidly grows complicated. If there are three red cards, for example, no one will react in the first round, which tells students that there must be more than 1 red card. When no one reacts in the second round, that tells students there must be more than two red cards. Then, those with red cards will know there are exactly three, and they will know their color.

Round 6 (liar's version)

What to do: Give out one red card and the rest black. There will be one student selected secretly to be a "liar."

Instruction to students: There is <u>at least</u> one red card. Students should close their eyes while I walk around and touch one person on the back. This person is the liar. When I ask whether you know your card color, you *must* lie. All others must tell the truth. Then: Raise your hand if you know your color. (Give students a moment to think about it, then have them put their hands down.) Now, raise your hand if you know who the liar is.

Analysis: The student with the red card should be the only one to raise his hand at first if everyone were telling the truth. So the liar will either be the student with the red card NOT raising his hand, or a student with a black card who IS raising his hand.

Extension: You can do different versions of this game with multiple liars and multiple red cards. Some of the versions may not be solvable, but as long as students are able to articulate why, they are getting something valuable out of the game.

Magic Squares Activity

This activity uses a lot of different mathematical concepts, including some of the logic concepts that have been touched on in this book. On the most basic level students are doing simple single-digit *addition* and *subtraction* and are using *deductive reasoning*. Older students will also use *division* and the concept of *averaging numbers* to reason about the magic squares. In the second part of the activity, students use the geometric concepts of *reflections*, *rotations*, and *symmetries of the square*. In the Fifteen Game, students use the concept of a *strategy*, and in relating the two games, students will see another example of an *isomorphism*.

Phase 1: Exhibition

Make an array on the board with the numbers 1 through 9.

1	2	3
4	5	6
7	8	9

Ask students to calculate the sum of each row, and record their answers on the board.

$$1 + 2 + 3 = 6$$
$$4 + 5 + 6 = 15$$
$$7 + 8 + 9 = 24$$

Once it is clear to students how to lay out the cards and what is meant by "calculating the sum of each row," they are ready for the puzzle.

Phase 2: Students Begin by Making an Array in Which the Sums of the Rows Are Equal.

Break students into groups of two or three, and give each group a set of cards, ace through 9. (The ace represents the 1.)

Ask students to make their own three-by-three array of cards.

Ask students to switch cards around until they end up with three rows that have exactly the same sum. (Do not tell students what the "target sum" is, only that it should be the same for each row.)

Most students eventually arrive at a solution. (There are several solutions.) If they need a hint, ask them to lay out three arbitrary rows, and take their sums. They can "fatten up" a lean sum, and "slim down" a fat sum, by trading a larger card for a smaller. In trading around like this, they ultimately solve the problem.

Students will complete this task at different rates. As individuals complete it, you can give them the next task.

Phase 3: Students Rearrange Their Arrays Until the Columns Also Have Equal Sums.

Students now have a three-by-three array of cards in which the rows all have the same sum. Ask them to "improve" the array, by making the *columns* also have the same sum.

Younger students often try to start the puzzle from scratch at this point, trying to come up with a new arrangement that has both row and column sums equal. Tell students that it is important for them to find a way to "fix" the columns without undoing the work they have already done on the rows. If they need a hint (and many do), show them that they can *keep the row sums the same, but change the column sums*, by trading cards within a row. Again, they can make lean columns fatter and fat columns leaner by trading within a row.

Students will finish this task at different rates. As they do, give them the next assignment.

Caution: students by now will have guessed that the common sum (the "magic constant") must be 15. It may happen that one of their columns of cards already sums to 15, and they will want to keep this column stable. But there are situations when the other two columns cannot balance while the stable column has a certain set of three cards. (It turns out that the puzzle cannot be solved if one of the columns contains 2, 5, 8 or 4, 5, 6. These are precisely the sets of numbers that will end up in the diagonals of the solved puzzle.) Students will have to break up the column that they think is already solved to get a sum of 15 for all three. They may need a hint if they are stuck here.

Phase 4: Students Rearrange Their Arrays Until the Diagonals Also Have Equal Sums

Another hint may be in order: if you take the top row, and make it the bottom row (this is physically awkward but conceptually easy), you haven't changed row or column sums, but you've changed the sums of the diagonals.

You can thus sort the rows. And you can do the same with the columns. Using this (not very obvious) procedure, you never need more than two steps to finish the puzzle.

If students are stuck here, an additional hint is to ask students to think about what number should go in the middle of the puzzle and then switch rows and columns around with that goal in mind. Many students will conclude that 5 should be in the middle, even without being able to explain why. (In fact, 5 is the median and the mean of the numbers, but the reason 5 has to be in the middle is more complex than that.)

As students finish the puzzle, have them put their results on the board for all to see. Each solution is what is classically called a "magic square."

Solutions:

2	9	4
7	5	3
6	1	8

6	7	2
1	5	9
8	3	4

8	1	6
3	5	7
4	9	2

4	3	8
9	5	1
2	7	6

6	1	8
7	5	3
2	9	4

8	3	4
1	5	9
6	7	2

4	9	2
3	5	7
8	1	6

2	7	6
9	5	1
4	3	8

Notice that the four solutions in the top row are all rotations of each other. The solutions in the bottom row can be generated by reflecting the top solutions about a horizontal line. There are other ways to reflect and rotate, but you will always end up with one of these eight solutions.

The instructor should allow the pairs of students to work, each at their own pace, until the majority of the class reaches a solution to this basic version of the puzzle.

An Extension for Students Who Work Ahead of the Class

Take away the ace and replace it with a 10. Now the students have the numbers 2 through 10. Ask the students to solve this variation of the puzzle. This has the effect of adding 9 to the total sum, adding 3 to each row, or adding 1 to each individual card. Most of the time, students start afresh with this problem, and go through the steps they used to get the original magic square, sometimes making the connection only after they have worked the problem the long way. (This is a fine way to build understanding.) Sometimes students do have the insight that they need only add 1 to each entry in the original square and then solve the problem quickly.

If students don't eventually make the connection themselves, put up a copy of the new (2 through10) magic square. Then subtract one from each entry, and let them discover that this gives them one of their old magic squares. The generalization then comes quickly.

To clinch it, you might ask what to do if students want to make a magic square of the numbers 4, 5, 6, 7, 8, 9, 10, 11, 12. (Add 3 to each entry of the original square.)

Once students have the idea of getting new squares from old, you can get them to focus on the process, rather than on the squares themselves. Multiply each entry of the original square (1 through 9) by 100. That is, simply place two zeroes after each entry. They will easily see that this new square is still magic.

It is sometimes useful *not* to describe the situation by saying that each number has been multiplied by 100. Instead, give students the numbers 3, 6, 9, 12, . . . 27 and ask them to form a magic square. (By this time, it is best to do things by pencil, rather than with cards.) Students will eventually see that you can simply multiply the original 1-through-9 square by 3. At that point, you can point out to students that the same process had been used in adjoining two zeroes to each entry earlier: you had multiplied by 100.

The next step, which your group may or may not be ready for, is to ask about a magic square using {4, 7, 10, 13, 16, 19, 22, 25, 28}. If the original square is denoted by S, we can call this square 3S + 1. Then you can give them a square with entries of the form 10S + 2 and challenge them to "decompose" it. That is, to explain how to get it from the original square.

What is going on here is that three-by-three magic squares form what is called a "vector space." If you drop the condition that the entries be distinct numbers, the three-by-three square consisting entirely of 1's (or 2's or 3's or 17's) counts as magic. Then, if S and T are magic squares, you can add them term-by-term to get a magic square S + T. Or, you can scale them: multiply each entry by a number k to get a magic square kS.

For example:

$$\begin{array}{|c|c|c|}\hline 2 & 9 & 4 \\ \hline 7 & 5 & 3 \\ \hline 6 & 1 & 8 \\ \hline \end{array} + \begin{array}{|c|c|c|}\hline 1 & 1 & 1 \\ \hline 1 & 1 & 1 \\ \hline 1 & 1 & 1 \\ \hline \end{array} = \begin{array}{|c|c|c|}\hline 3 & 10 & 5 \\ \hline 8 & 6 & 4 \\ \hline 7 & 2 & 9 \\ \hline \end{array}$$

$$3 \times \begin{array}{|c|c|c|}\hline 2 & 9 & 4 \\ \hline 7 & 5 & 3 \\ \hline 6 & 1 & 8 \\ \hline \end{array} = \begin{array}{|c|c|c|}\hline 6 & 27 & 12 \\ \hline 21 & 15 & 9 \\ \hline 18 & 3 & 24 \\ \hline \end{array}$$

$$\begin{array}{|c|c|c|}\hline 2 & 9 & 4 \\ \hline 7 & 5 & 3 \\ \hline 6 & 1 & 8 \\ \hline \end{array} + \begin{array}{|c|c|c|}\hline 7 & 8 & 3 \\ \hline 2 & 6 & 10 \\ \hline 9 & 4 & 5 \\ \hline \end{array} = \begin{array}{|c|c|c|}\hline 9 & 17 & 7 \\ \hline 9 & 11 & 13 \\ \hline 15 & 5 & 13 \\ \hline \end{array}$$

The structure of a vector space is one that is basic to much mathematics, and this activity is a springboard to the notion of a linear combination, of a basis, and of much more that students may discover in more advanced work. But for now, the insight that one magic square generates many others is more than enough.

Phase 5: Whole-Group Instruction

(1) Choose three solutions.

Students will have written their solutions on the board. Choose and circle three different solutions. Among the three you choose, try to get two solutions that differ by a line reflection, and two that differ by rotation.

(2) Help students see how the solutions are related.

Let the students observe the three solutions you've chosen, and ask if they are really all different. They are likely to see so for themselves, especially if you've been able to choose two solutions that differ by reflection.

The topic now becomes geometry, rather than arithmetic. There are eight possible arrangements of the numbers 1 through 9 in a magic square, corresponding to the eight symmetries of a square.

That is, from any one solution, you can get to any other by rotation (four rotations, including a rotation through zero degrees) or line reflection (in the two diagonals and in a horizontal and vertical midline of the square). Students generally understand that you can reflect a solution to get another solution. The fact that they can rotate solutions is a bit more difficult, but usually arises by inspecting different solutions.

(3) Ask students to find all possible solutions.

Ask students how many ways you can take one solution and get others. There are eight ways, including "doing nothing." That is, there are eight symmetries of a square. Introduce the term "symmetry," a way to move a figure so that it coincides with itself. Don't forget to count the "identity" symmetry, which is equivalent to a rotation by zero degrees.

To clinch the idea, you can ask students how many symmetries an equilateral triangle has. There are six symmetries: three rotations and three line reflections. But they are harder to spot than the symmetries of a square. The reflections are in oblique lines and are hard to distinguish from rotations.

(4) Advanced topic: Combining symmetries

You now have three solutions on the board, say square A, square B, and square C. Have students identify the rotation or reflection that takes A onto B, and also B onto C. Then ask them if they can describe a motion that takes A directly onto C, without going "through" B. They usually can do this. See the example below, in which rotating and then reflecting the square about a horizontal line gives us the same solution as reflecting it about a diagonal line (here, the diagonal that extends from bottom left to top right). Notice that only the corner numbers are given because they uniquely determine the placement of the other numbers.

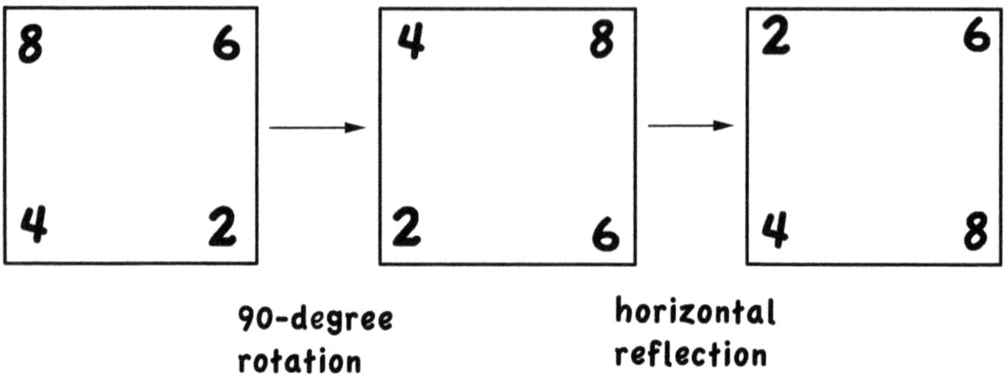

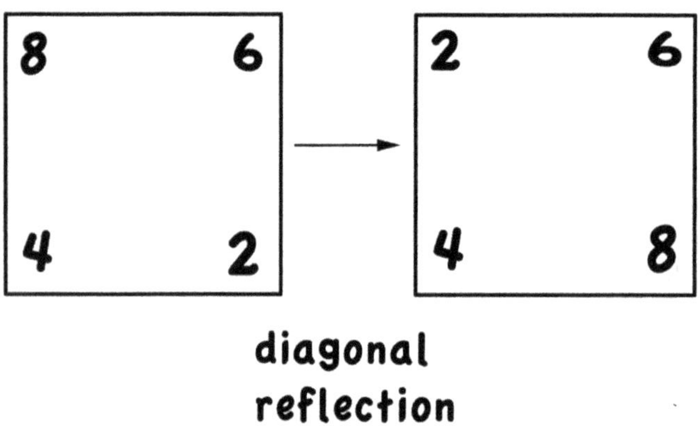

This question leads to the algebra of symmetries: you can perform one symmetry, then the next, and the result is a third symmetry. This is called "composition" of symmetries.

Next insight, for advanced students: Combining symmetries is not commutative (!). Suppose we have symmetry (a) that takes square A into square B, and symmetry (b) that takes square B into square C:

A ----(a)----->B   B ----(b)----->C

Then the composition a*b, takes A directly into C:

A -----(a*b)-----> C .

But suppose we compose in the other direction? What does b*a do to square A:

A------(b*a)----->???

In general, students will find that the result is not square C, but some other position of the magic square. It makes a difference, usually, which symmetry you apply first, and which you apply second. Composition of symmetries is not *commutative.*

Note: it is vital, for this step, that (a) and (b) not both be rotations. Rotations will commute with each other. But if at least one of (a) or (b) is a line reflection, the composition of those two symmetries will not commute.

In advanced work, mathematicians study the *dihedral group* of symmetries of a square. This is the set of eight symmetries, together with the table of operations that they generate. While it is well within the abilities of middle school students to generate this table, and to make some interesting deductions and observations, we make this note simply to observe that the mathematics here is quite serious.

The Fifteen Game

This is a simple game and is worthwhile here because it relates nicely to the magic squares. Students will get a feel for this relationship, and the instructor can help them make this thought explicit. It is best to leave a little time (perhaps a day) between the Magic Square activity and the Fifteen Game so that the relationship is not too obvious.

Phase 1: Exhibition

Ask for a volunteer to play against the instructor. As always, this stage is meant simply to clarify the rules. The instructor should let the student win.

The Rules:

Write the numbers 1 through 9 on the board in a straight line.

The first person "takes" a number by crossing it out and rewriting it on his side of the board.

The winner is the first person to make a sum of 15 with exactly three cards. (9 + 6 doesn't work because those are only two cards.) In the example below, Katie has won the game because she has 3, 4, and 8.

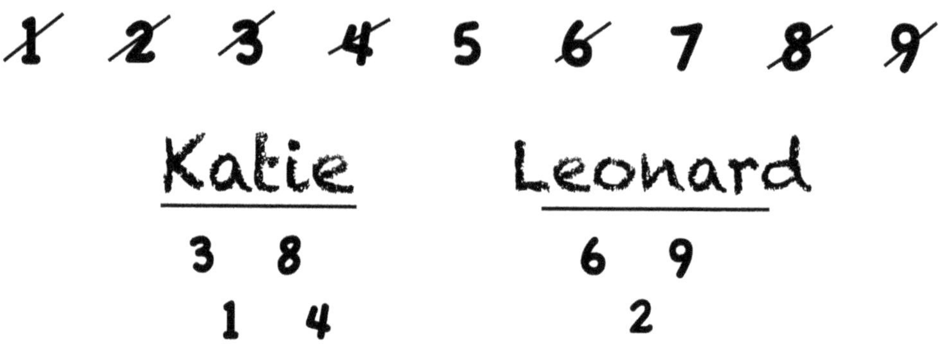

The rules of this game are hard to describe, and the strategy harder to see, than in previous games. Students will think that two cards (say, 9 and 6) constitute a win, because they add up to 15. Or, they will think that once a player has chosen three cards (say, 3, 5, and 6) he has lost, because they don't add up to 15.

As usual, it does not help to try to clarify the rules in advance. They become clear during the play. In fact, part of the reason to play exhibition games is to elucidate the rules, and part of the learning for the students is to understand the rules, and then their consequences. ("Rules and consequences" imitates the structure of axioms and theorems in more serious mathematics.)

Phase 2: Students Play the Game and Look for a Strategy.

Break students into pairs and give each pair the cards ace (1) through 9.

Instruct students to lay out the nine cards in a row on the table and play the game. Be sure that students are not hiding cards from each other once they have taken a card. All cards should be visible.

It is unlikely that they will develop a reliable strategy, but they may get a "feel" for it.

In fact, there is no winning strategy for either player. Perfect play results in a tie game. You will know that students understand the game when they start tying very often.

It is best not to talk much about a strategy for this game. Students will understand when we relate the game to the magic square puzzle.

In fact, this game is isomorphic to tic-tac-toe, played on a magic square. But students can have the joy of discovering this for themselves. Many will comment, as they play, that the strategy is to block one's opponent, and that it feels like tic-tac-toe. It is best, usually, to share this comment, but not to make it precise. The isomorphism, revealed later, provides precision.

Phase 3: Class Discussion in Which Students Relate the Two Games

Call students back together to discuss the Fifteen Game.

Set up a Fifteen Game and a solved magic square on the board.

Have students play an exhibition game of Fifteen again on the board. This time keep track of the moves by placing an x (for one player) and an o (for another player) on the numbers of a magic square. Students will see, usually to their delight, that they've been playing tic-tac-toe.

Their idea of a blocking strategy is exactly what they know from that game. They had been intuiting the isomorphism before it was made precise in this way.

Questions and Answers

Q. What is the purpose of the activities in this book?

A. This book is not a course in mathematics or even a course in logic. Instead, it is meant to give teachers and parents ideas for activities that will tap into kids' intuitions about logical reasoning and give them practice with it. Students will also gain an appreciation of the power of logic as they use it to reach non-intuitive results.

Q. Is it important to do the activities in order?

A. The activities are modular, and the instructor should feel free to change the order of the modules in any way that makes sense. The order presented, however, has worked well, and thought was given to mixing up types of activities so that the children remain engaged and interested.

Q. The activities do not include guides for timing. How long should I spend on each one?

A. We recommend that a teacher stick with an activity as long as the students are engaged, which can vary depending on the age of the children, the ability level of the children, the time of day, and other such factors.

The content of the lesson is the thinking itself, rather than any particular mathematical result. So as long as the children are receptive to exploring more questions, the teacher should continue. If it takes an hour to explore a single question because the students are analyzing it deeply, all the better. If, on the other hand, students grow restless after twenty minutes, it many be time to take a break or switch to a different activity and try the first one again another day. In fact, many of the activities are difficult to explore fully in one session and should be revisited two or three times.

Q. How can I accommodate students who are working at different paces?

A. 'When students are given problems requiring thought that allow for multiple solutions or different approaches, they will naturally work at very different paces. So the difficulty that arises should not be seen as a problem but rather a sign that the instructor is doing something right.

Still, allowing students to work at their own pace certainly does make it more difficult to manage a class. In general, the activities in this book are structured to allow a balance. Each activity starts off in a structured way with a few carefully chosen examples so that the students have just enough information to dig into the problems. In this initial stage, all students are part of the same conversation, and all are starting in the same place.

Afterwards students frequently go off on their own or in groups to work through examples and make their own discoveries. During this stage the teacher's role changes. When things are going well, the teacher is merely a "fly on the wall." When a child disengages, runs out of ideas, or gets frustrated, then it is time for the teacher to step in and help the student reengage – by asking a question, posing a challenge, or clearing up some point of confusion. During this period students will work at different paces.

In the final phase of an activity, it is helpful for the group to come back together as a whole and share their ideas and discoveries with each other. This is important because the students learn from each other, and it is especially true with problems that have multiple approaches or multiple solutions. Additionally, it is important for students to articulate and defend their positions. A student may have a hunch that a letter's value is 5 in one of the cryptarithms and may even have stumbled upon the solution by accident, but explaining why the solution must be five helps him to clarify the logical steps involved and brings him to a higher level of understanding.

Q. What should I do to prepare for a lesson?

A. The best way to prepare to lead students through an activity is simply to work through the problems. Remember that the lessons are not meant to be a static step-by-step script of what to say to a class but rather a springboard for exploring interesting questions.

By working through the problems, the instructor will naturally have questions and moments of confusion – perhaps the same ones that the students will have. By going through this process beforehand, the instructor will be better prepared to guide the students through their own exploration.

Even with careful preparation, an instructor cannot know everything and cannot predict what will happen. The best instructors are comfortable not knowing everything and are willing to learn about the problems from their students.

To prepare to lead a post-activity class discussion, read through the lesson plans and then observe the students carefully as they work through the problems. Try to pull out a few "big ideas" from the activity that you may wish to highlight in the discussion. These themes may come from the lesson plans themselves, or they may emerge as the students work through the problems. For example, a teacher may want to call attention to a particular problem-solving strategy that students used successfully, or he may want to highlight an interesting connection to another problem.

Q. How can I adapt the lessons for use with different age groups or different ability levels?

Most of the lessons start of with a simple question, and then the material gets progressively deeper and more difficult. Typically, stronger students will simply move faster and further through the material. For very young students, two easy cryptarithms may be enough to fuel forty-five minutes of exploration and discussion, whereas older students may see those answers right away and linger instead over the challenge cryptarithms.

Q. How much math background do I need to conduct these lessons?

A. The activities here require almost no formal background in mathematics. There are times when technical vocabulary is used, but even then no prior knowledge is assumed. Both the instructor and the students will gain a fuller understanding of the mathematical terms as they work through the activities.

The only real math requirement to facilitate these activities is comfort and fluency with basic arithmetic. Additionally, there are some optional extension activities in which a background in mathematics would help an instructor to lead a richer exploration and discussion with highly advanced students. But the activities are rich even without those extensions.

We have found that the most important qualities for an instructor are the same ones we want to cultivate in our students: the qualities of courage, persistence, and joy in solving problems and puzzles.

The instructors who find the most success with these materials are those who feel comfortable asking and being asked questions that they don't yet have an answer to.

Glossary

Contradiction

A *contradiction* is a statement that is always false, no matter what observations of reality we might make. For example, "The sky is blue and also not blue." No matter what color the sky actually is, this statement must be false.

One type of mathematical proof is a "proof by contradiction." In such a proof we establish a statement of the form, "If A is true, then A is also false." If we know this, then A cannot be true.

In other arguments, we look at two statements, P, Q, and form the statement "P and Q." If this statement is always false, then either P or Q must be false. They cannot simultaneously be true.

Cryptarithm

A *cryptarithm* is a classic type of mathematical puzzle. In a cryptarhythm, an equation is given in which the digits are replaced by letters. The same digit is always replaced by the same letter, and two different digits must be replaced by two different letters. The goal is to figure out which digit each letter stands for.

A cryptarithm is considered especially elegant if it has a unique solution.

Euler Path

A *Euler path* is a path through a graph that traces every edge exactly once.

Graph Theory

Graph theory is the study of *graphs*. In this context, a graph is a collection of points (*vertices*), some of which are connected by curves or line segments (*edges*). The graph below, for example, is made of four vertices (A, B, C, and D), and four edges connecting them.

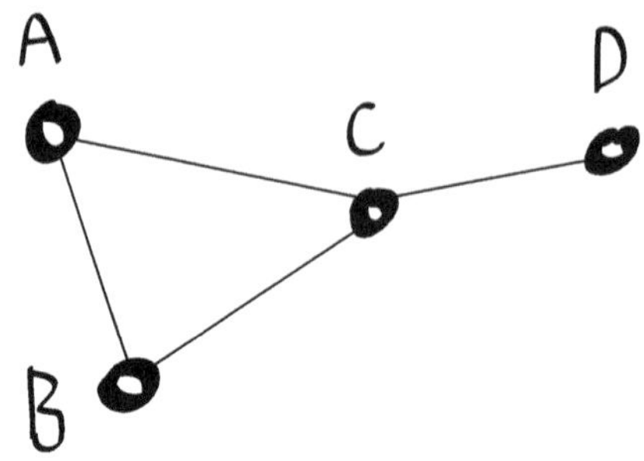

Graphs are useful to model relationships. For example, the graph below the graph might represent friendships among four people. It then would show that C is a friend to everyone, while D has only one friend. The same graph could also represent four towns connected by roads. (Note that the graph does not give all the information that a map would: the distances are not in proportion and the shapes of the roads are not given by the edges.) Or, it could show which four dishes on a menu share common ingredients. (Dish A shares an ingredient with B and C, but it has nothing in common with D.)

Implication

An *implication* is a relationship between two statements. Formally, if the statements are A and B, then an implication is the statement "If A then B", or "A implies B". The precise definition of the statement is not easy to understand, but for our purposes it is enough to think that it means that if A is true, then B is also true.

Implication lies at the heart of logic and, in turn, of mathematics. Implication is how we derive new truths from the ones that are already established or assumed.

Invariant

An *invariant* is a characteristic of a mathematical object that remains unchanged after a transformation. For example, when triangle is scaled up in size, the interior angles remain the same (although the side lengths do not). So the measures of the internal angles of a triangle are *invariant* with respect to scaling of the triangle.

Or if we slide each point of a line to the right three inches, the locations of various points change. But the distance between two points is *invariant.*

Isosorphism

"Isomorphism" comes from the Greek words meaning "same" and "shape." An *isomorphism* is a way of pairing up elements of two mathematical objects, so that a common structure is preserved.

For example, the experiment of flipping two coins is isomorphic to drawing either a red or black ball from a bag containing one of each, replacing it, and then picking again. The outcome of the flip of one coin can be paired with the outcome of drawing out the first ball (and noting its color). The outcome of the flip of the other coin can be paired with the outcome of drawing out the second ball. The two situations can be modeled with the same sample space, and the same probability tree.

Objects for which we can find an isomorphism are called *isomorphic*. Two objects that are isomorphic may look superficially different, but can be analyzed in exactly the same way: they have the same structure.

Parity

Every integer is either even or odd. This is its *parity*. For example, the parity of 3 is odd, and the parity of 28 is even. Sometimes we can tell whether the parity of a number is odd or even, despite the fact that we don't know what the number is. This possibility makes parity a powerful tool in mathematical proofs. For example, suppose a farmer is making a rectangular enclosure out of meter-long lengths of fence. Then we can tell for sure that he will need an even number of fence lengths, even though we may not know how large the enclosure will be. (Do you see why?)

Pigeonhole Principle

The *pigeonhole principle* asserts that if we must classify n+1 objects into n categories, then two of the objects must go in the same category. The name comes from a classic situation: if there are n+1 pigeons nesting in n "pigeonholes" (or nesting boxes), then one pigeonhole must contain at least two pigeons.

The principle may seem trivial or obvious, but it is very powerful, and can be applied to an enormous range of problems often leading to surprising results.

Proof

A *mathematical proof* is a chain of implications leading from one set of statements (the "given" statements) to another. If the first set of statements is true, we say that the second set is *proved* by the existence of this chain of implications.

In mathematics, unlike the sciences, no reference is made to empirical evidence or external facts in establishing the truth of the statement. Instead, a mathematical proof rests solely on initial assumptions (called "axioms"), on the given statements, and on statements that have been previously proved.

Systems of Numeration

A *system of numeration* is a way of expressing numbers in writing. An example is Roman numerals (MMXIV). The Mayans used dots and lines. The one we most commonly use (with the digits 0 through 9) is a *positional* system: the value of a digit depends on its position within the number. For example, the 7 in 73 means something different from the 7 in 7000 because they are in different positions. We say our number system is a "base 10" system because the value of the positions is based on the powers of 10: 1, 10, 100, 1000, etc. Other positional systems can have other bases. A base 3 number system has the following place values: 1, 3, 9, 27, 81 . . . . In general, a *base b number system* is a positional number system based on the powers of b, where b is a whole number greater than 1.

Venn Diagram

A *Venn diagram* is a model for showing the relations between sets of objects. The sets are often represented with circles or ovals, with overlapping circles representing the fact that some objects belong to both sets.

www.ingramcontent.com/pod-product-compliance
Ingram Content Group UK Ltd.
Pitfield, Milton Keynes, MK11 3LW, UK
UKHW051452290926
15471UKWH00008B/1166

9 780977 69396